$4

Audio Systems Technology

LEVEL I
Handbook for Installers and Engineers

Audio Systems Technology

LEVEL I
Handbook for Installers and Engineers

Written by Larry W. Garter, CET
Additional Material Provided by R. David Reed

National Systems Contractors Association

PUBLICATIONS

PROMPT® Publications is an imprint of Howard W. Sams & Company, A Bell Atlantic Company, 2647 Waterfront Parkway, E. Dr., Indianapolis, IN 46214-2041.

International Standard Book Number: 0-7906-1162-7
Library of Congress Catalog Card Number: 98-67126

Acquisitions Editor: Candace Hall, Loretta Yates
Editor: Natalie F. Harris
Assistant Editors: Pat Brady, J.B. Hall
Typesetting: Natalie Harris
Indexing: Pat Brady, Natalie Harris
Cover Design: Kelli Ternet
Graphics: Supplied by NSCA
Graphics Conversion: Kelli Ternet, Terry Varvel

PRINTED IN THE UNITED STATES OF AMERICA

9 8 7 6 5 4 3

Contents

Chapter 3
System Elements   *131*

Appendix A
Answers to Chapter Questions   *259*

Dedication

NSCA gratefully dedicates this book to Jay Johnson, longtime NSCA board member and audio industry leader. Jay's vision of a highly professional, certified technical audio work force is being realized due to his leadership and dedication. This book is his legacy; it is both a culmination of his efforts to create an audio technician certification program, and the beginning of a new era of training and education that will raise the standard for all audio professionals.

Preface
How to Use This Book

This book is part of the NSCA "Audio Technology Handbook" series. The series consists of three books, each corresponding to a different level of NICET certification. (The Level III book is in progress.) Level I is essentially entry-level material best suited to the technician trainee, or for use as a review by experienced technicians preparing for NICET. Level II is for the installer or an associate's degree or equivalent. Level III, when it becomes available, will be for the more advanced technician who has several years of job experience and an associate's degree or equivalent.

While each book in the series contains its own unique information, there is also a bit of overlap from one level to the next. This intentional dovetailing provides repetition on the most important, fundamental points, and allows the entire series to be used as a systematic, progressive course of study from the basics up to the intermediate and advanced levels.

This book is meant to be used in a variety of ways: as a study and preparation guide for those seeking NICET certification in audio systems; as a training manual for installers and designers of audio systems; as an overview of the fundamental things a good audio technician should know; and as a reference guide for installers and engineers needing a one-stop source of practical technical audio information.

The content of the chapters is organized loosely on "work elements," the subjects encountered on the NICET exams. All of the "core" elements and most of the general work elements are represented here. Each chapter begins by stating the NICET work element descriptions that apply to the chapter's content. Because many of NICET's special work elements deal with special systems or subjects that are not universal to the present-day audio industry, and because they are elective on NICET exams, we have chosen not to discuss them at length in this book. A list of work elements not discussed can be found in the appendix of this book. If you are not familiar with NICET's audio technician certification program, you can read about it at the end of the book, in Appendix B, "The NICET Audio Certification Program."

Each chapter concludes with review questions designed to encourage readers to test their understanding of the chapter content by attempting to answer them. Answers and solutions to all review questions are listed in Appendix A in the back of the book. Trainers may want to use the review questions as homework, or as an aid for live instruction. For those preparing for the NICET exams, the questions are an excellent way to get familiar with NICET-type exam questions.

Acknowledgments

Again, NSCA gratefully dedicates this book to Jay Johnson, longtime NSCA board member and audio industry leader. This book is his legacy; it is both a culmination of his efforts to create an audio technician certification program, and the beginning of new era of training and education that will raise the standard for all audio professionals.

We would also like to thank the following people who played a major role in making this book possible: Bob Bushnell, Jim Brawley, Larry Garter, Ian Wolfe, Karen Hunt, Melvin Wierenga, Matthew Marth, Ted Uzzle, Bill Whitlock, Candace Drake Hall.

NSCA is grateful to the following individuals for reviewing and making suggestions for this book: James Beckham, Bob Bushnell, Larry Garter, Per Forsberg, John Lanphere, Matt Marth, Gary Schmitt, and Ian Wolfe.

Chapter 1
Communications

- **NICET Work Element 11001, Basic Communication Skills**
- **NICET's Description:** "Use proper punctuation, vocabulary, spelling, and sentence structure.
  Follow written instructions. (See dictionaries and basic grammar references.)"
- **NICET Work Element 11006, Basic Plans and Specifications**
- **NICET's Description:** "Recognize, or determine from symbol lists, symbols for basic sound system components (HVAC and electrical fixtures/devices and symbols for basic structural components. Use a scale or ruler to measure architectural dimensions and determine equipment locations."

Chapter Contents

BASIC COMMUNICATION SKILLS
Introduction
Words Often Misspelled
Words and Expressions Commonly Misused
Elementary Rules of Usage
Other References

BASIC PLANS and SPECIFICATIONS
Introduction
How are Plans and Specifications Prepared?
An Architect Does Not Have an Easy Task
Divisions of the Plans and Specifications
Plan Scaling
Elements of a Specification
Specification Sections
Questions

BASIC COMMUNICATIONS SKILLS

Introduction

When we speak of effective communications skills, we are allud-
ing to a person's ability to make an intelligent presentation; that the
listeners are able to understand clearly what the talker is attempting to convey. Despite the relative merit or importance of the presentation, a spoken or written communication which is couched in
poor grammar, illogically presented, or which uses improperly applied vocabulary is not going to receive the attention that the subject might otherwise warrant. It has been observed that 74% of
businesses in the United States cite poor communications skills as
one of the most serious and disruptive influences in their conduct
of day-to-day business activities. Anyone who has aspirations to
advance in their chosen profession must learn to express themselves
effectively in both spoken and written communications.

From what has been said above, it should be obvious that when we
attempt to communicate it is imperative that we use correct grammar and precise vocabulary. The likelihood of advancing in any
organization is highly dependent on being able to express one's
self in a clear and well organized manner. Good diction and the
choice of proper words to describe what you are trying to communicate will go a long way toward avoiding misunderstandings.

Unfortunately, there is a general lack of emphasis on teaching proper
English grammar and speech in United States high schools. The
use of improper grammar is not confined to high-school dropouts.
We have probably all met a few college graduates whose communications skills have barely advanced beyond the 5th-grade level.

Computer programs can check spelling, and there are some that
will analyze the grammatical structure of your written word. However, computers are completely lacking in common sense and they
won't tell whether you have used the right wording or the proper

nouns to describe a situation which you are attempting to convey to others.

Words Often Misspelled

The following words are commonly misspelled and should be noted for future reference.[1] Most people have their own set of commonly misspelled words. You should be aware of yours and make every effort to check your spelling, as nothing stands out more than a misspelled word.

accidentally
advice
affect
beginning
believe
benefit
challenge
criticize
deceive
definite
describe
despise
develop
disappoint
duel
ecstasy
effect
existence
fiery
formerly
humorous
hypocrisy
immediately
incidentally
latter

led
lose
marriage
mischief
murmur
necessary
occurred
parallel
Philip
playwright
preceding
prejudice
principal
privilege
pursue
repetition
rhyme
rhythm
ridiculous
sacrilegious
seize
separate
shepherd
siege
similar

simile
too
tragedy
tries
undoubtedly
until

Words and Expressions Commonly Misused

Many of the words and expressions discussed here are not so much bad English as bad style. The proper correction is likely to be not the replacement of one word or set of words by another, but the replacement of vague generality by definite statement. The following is just a sample of the words and expressions commonly misused.

All right. Idiomatic in familiar speech as a detached phrase in the sense, "Agreed," or "Go ahead." In other uses better avoided. Always written as two words.

As good or better than. Expressions of this type should be corrected by rearranging the sentence:

Undesirable: *My opinion is as good or better than his.*

Better: *My opinion is as good as his, or better (if not better).*

As to whether. "Whether" is sufficient; leave out the "as to" part.

Case. The Concise Oxford Dictionary begins its definition of this word: "instance of a thing's occurring; usual state of affairs." In these two senses, the word is usually unnecessary:

Undesirable: *In many cases, the rooms were poorly ventilated.*

Better: *Many of the rooms were poorly ventilated.*

Undesirable: *It has rarely been the case that any mistake has been made.*

Better: *Few mistakes have been made.*

(See Wood, Suggestions to Authors, pp. 68-71, and Quiller-Couch, *The Art of Writing*, pp. 103-106.)

Certainly. Used indiscriminately by some speakers, much as others use *very*, to intensify any and every statement. A mannerism of this kind, bad in speech, is even worse in writing.

Compare. To "compare" to is to point out or imply resemblances between objects regarded as essentially of different order; to compare with is mainly to point out differences between objects regarded as essentially of the same order. Thus, life has been compared to pilgrimage, to a drama, to a battle; Congress may be compared with the British Parliament. Paris has been compared to ancient Athens; it may be compared with modern London.

Consider. Not followed by "as" when it means, "believe to be." "I consider him thoroughly competent." Compare, "The lecturer considered Cromwell first as soldier and second as administrator," where "considered" means "examined" or "discussed."

Dependable. A needless substitute for *reliable, trustworthy*.

Due to. Incorrectly used for *through*, *because of*, or *owing to*, in verbal phrases: "He lost the first game due to carelessness." In correct use related as predicate or as modifier to a particular noun: "This invention is due to Edison;" "losses due to preventable fires."

Effect. As noun, means "result"; as verb, means "to bring about," "accomplish" (not to be confused with *affect*, which means "to influence"). As a noun, often loosely used in perfunctory writing about fashions, music, painting, and other arts: "an Oriental effect;" "effects in pale green;" "very delicate effects;" "broad effects;" "subtle effects;" "a charming effect was produced by." The writer who has definite meaning to express will not take refuge in such vagueness.

Etc. Not to be used of persons. Equivalent to "and the rest," "and so forth," and hence not to be used if one of these would be insufficient, that is, if the reader would be left in doubt as to any important particulars. Least open to objection when it represents the last terms of a list already given in full, or immaterial words at the end of a quotation. At the end of a list introduced by "such as," "for example," or any similar expression, *etc.* is incorrect.

Fact. Use this word only of matters of a kind capable of direct verification, not of matters of judgment. That a particular event happened on a given date, or that lead melts at a certain temperature, are facts. But conclusions such as that Napoleon was the greatest of modern generals, or that the climate of California is delightful, however incontestable they may be, are not properly facts.

Factor. A hackneyed word; the expressions of which it forms part can usually be replaced by something more direct and idiomatic.

Undesirable: *His superior training was the great factor in his winning the match.*

Better: *He won the match by being better trained.*

Undesirable: *Heavy artillery is becoming an increasingly important factor in deciding battles.*

Better: *Heavy artillery is playing a larger and larger part in deciding battles.*

Fix. Colloquial in America for "arrange," "prepare," "mend." In writing restrict it to its literary senses, "fasten," "make firm or immovable," etc.

However. In the meaning "nevertheless," not to come first in its sentence or clause.

Undesirable: *The roads were almost impassable. However, we at last succeeded in reaching camp.*

Better: *The roads were almost impassable. At last, however we succeeded in reaching camp.*

When *however* comes first, it means "in whatever way" or "to whatever extent."

However you advise him, he will probably do as he thinks best.

However discouraging the prospect, he never lost heart.

Kind of. Not to be used as a substitute for "rather" (before adjectives and verbs), or except in familiar style, for "something like" (before nouns). Restrict it to its literal sense:

Amber is a kind of fossil resin.

I dislike that kind of notoriety.

The same holds true for *sort of.*

Less. Should not be misused for "fewer."

Undesirable: *He had less men than in the previous campaign.*

Better: *He had fewer men than in the previous campaign.*

Less refers to "quantity," *fewer* to "number." "His troubles are less than mine" means "His troubles are not so great as mine." "His troubles are fewer than mine" means "His troubles are not so numerous as mine." It is, however, correct to say, "The signers of the petition were less than a hundred," where the round number, a hundred, is something like a collective noun, and *less* is thought of as meaning "a less quantity or amount."

Line; along these lines. In the sense of "course of procedure," "conduct," "thought," *line* is allowable, but has been so much overworked, particularly in the phrase "along these lines," that a writer who aims at freshness or originality had better discard it entirely:

Undesirable: *Mr. B. also spoke along the same lines.*

Better: *Mr. B. also spoke to the same effect.*

Undesirable: *He is studying along the lines of French literature.*
Better: *He is studying French literature.*

Literal, literally. Often incorrectly used in support of exaggeration or metaphor.

Undesirable: *A literal flood of abuse.*

Better: *A flood of abuse.*

Undesirable: *Literally dead with fatigue.*

Better: *Almost dead with fatigue (dead tired).*

Lose out. Meant to be more emphatic than *lose*, but actually less so, because of its commonness. The same holds true of *try out*, *win out*, *sign up*, *register up*. With a number of verbs, *out* and *up* form idiomatic combinations: *find out*, *run out*, *turn out*, *cheer up*, *dry up*, *make up*, and others, each distinguishable in meaning from the simple verb. *Lose out* is not.

Most. Not to be used for *almost*:

Undesirable: *Most everybody*

Better: *Almost everybody*

Undesirable: *Most all the time*

Better: *Almost all the time*

People. The people is a political term, not to be confused with *the public*. From the people comes political support or opposition; from the public comes artistic appreciation or commercial patronage.

The word *people* is not to be used with words of number, in place of *persons*. If of six "people," five went away, how many "people" would be left?

Respective, respectively. These words may usually be omitted with advantage:

> Undesirable: *Works of fiction are listed under the names of their respective authors.*

> Better: *Works of fiction are listed under the names of their authors.*

In some kinds of formal writing, as in geometrical proofs, it may be necessary to use *respectively*, but it should not appear in writing on ordinary subjects.

So. Avoid, in writing, the use of *so* as an intensifier: "so good;" "so warm;" "so delightful."

State. Not to be used as a mere substitute for "say" or "remark." Restrict it to the sense of "express fully or clearly," as, "He refused to state his objections."

Thanking you in advance. This sounds as if the writer meant, "It will not be worth my while to write to you again." Simply write, "Thanking you," and if the favor which you have requested is granted, write a letter of acknowledgment.

They. A common inaccuracy is the use of the plural pronoun when the antecedent is a distributive expression such as *each, each one, everybody, every one, many a man,* which, though implying more than one person, requires the pronoun to be in the singular. Similar to this, but with even less justification, is the use of the plural pronoun with the antecedent *anybody, any one, somebody, some one,* the intention being either to avoid the awkward "he or she," or to avoid committing oneself to either. Some bashful speakers even say, "A friend of mine told me that they, etc."

Use *he* with all the above words, unless the antecedent is, or must be, feminine.

Very. Use this word sparingly. Where emphasis is necessary, use words strong in themselves.

While. Avoid the indiscriminate use of this word for "and," "but," and "although." Many writers use it frequently as a substitute for "and" or "but," either from a mere desire to vary the connective, or from uncertainty which of the two connectives is the more appropriate. In this use it is best replaced by a semicolon.

> Undesirable: *The office and salesroom are on the ground floor, while the rest of the building is devoted to manufacturing.*

> Better: *The office and salesroom are on the ground floor; the rest of the building is devoted to manufacturing.*

Its use as a virtual equivalent of "although" is allowable where this leads to no ambiguity or absurdity:

> *While I admire his energy, I wish it were employed in a better cause.*

This is entirely correct, as shown by the paraphrase:

> *I admire his energy; at the same time I wish it were employed in a better cause.*

Compare:

> *While the temperature reaches 90 or 95 degrees in the daytime, the nights are often chilly.*

Although the temperature reaches 90 or 95 degrees in the daytime, the nights are often chilly.

The paraphrase:

The temperature reaches 90 or 95 degrees in the daytime; at the same time the nights are often chilly.

— shows why the use of *while* is incorrect.

In general, the writer will do well to use *while* only with strict literalness, in the sense of "during the time that..."

Whom. Often incorrectly used for *who* before he said or similar expressions, when it is really the subject of a following verb:

Incorrect: *His brother, whom he said would send him the money...*

Correct: *His brother, who he said would send him the money...*

Incorrect: *The man whom he thought was his friend...*

Correct: *The man who (that) he thought was his friend (whom he thought his friend)...*

Would. A conditional statement in the first person requires *should*, not *would*:

I should not have succeeded without his help.

The equivalent of *shall* in indirect quotation after a verb in the past tense is *should*, not *would*:

He predicted that before long we should have a great surprise.

To express habitual or repeated action, the past tense, without *would*, is usually sufficient, and from its brevity, more emphatic:

Undesirable: *Once a year he would visit the old mansion.*

Better: *Once a year he visited the old mansion.*

Elementary Rules of Usage

1. Form the possessive singular of nouns with "'s." Follow this rule whatever the final consonant. Thus write:

    Charles's friend
    Burns's poems
    The witch's malice

    This is the usage of the United States Government Printing Office and of the Oxford University Press.

    Exceptions are the possessives of ancient proper names ending in *-es* and *-is*, the possessive Jesus,' and such forms as "for conscience' sake," "for righteousness' sake." But such forms as "Achilles' heel," "Moses' laws," "Isis' temple" are commonly replaced by:

    The heel of Achilles
    The laws of Moses
    The temple of Isis

    The pronominal possessives *hers*, *its*, *theirs*, *yours*, and *oneself* have no apostrophe.

2. In a series of three or more terms with a single conjunction, use a comma after each term except the last:

    Red, white, and blue.

Honest, energetic, but headstrong.

He opened the letter, read it, and made a note of its contents.

This is also the usage of the Government Printing Office and of the Oxford University Press.

In the names of business firms the last comma is omitted:

Brown, Shipley and Company

3. Enclose parenthetic expressions between commas.

The best way to see a country, unless you are pressed for time, is to travel on foot.

This rule is difficult to apply. It is frequently hard to decide whether a single word, such as "however", or a brief phrase, is or is not parenthetic. If the interruption to the flow of the sentence is slight, the writer may safely omit the commas. But whether the interruption be slight or considerable, never omit one comma and leave the other. Such punctuation as:

Marjorie's husband, Colonel Nelson paid us a visit yesterday.

— or:

My brother you will be pleased to hear, is now in perfect health.

— is indefensible.

Nonrestrictive relative clauses are, in accordance with this rule, set by commas:

The audience, which had at first been indifferent, became more and more interested.

Similar clauses introduced by "where" and "when" are similarly punctuated"

In 1769, when Napoleon was born, Corsica had but recently been acquired by France.

Nether Stowey, where Coleridge wrote *The Rime of the Ancient Mariner*, is a few miles from Bridgewater.

In these sentences the clauses introduced by *which*, *when*, and *where* are nonrestrictive. They do not limit the application of the words on which they depend, but add, parenthetically, statements supplementing those in the principal clauses. Each sentence is a combination of two statements which might have been made independently:

The audience was at first indifferent. Later it became more and more interested.

Napoleon was born in 1769. At that time Corsica had but recently been acquired by France.

Coleridge wrote *The Rime of the Ancient Mariner* at Nether Stowey. Nether Stowey is only a few miles from Bridgewater.

Restrictive relative clauses are not set off by commas:

The candidate who best meets these requirements will obtain the place.

In this sentence the relative clause restricts the application of the word *candidate* to a single person. Unlike those above, the sentence cannot be split into two independent statements.

4. The abbreviations *etc.* and *jr.* are always preceded by a comma, and except at the end of a sentence, followed by one.

    Similar in principle to enclosing parenthetic expressions between commas, is setting off phrases or dependent clauses preceding or following the main clause of a sentence by commas. The sentences quoted in this section and under rules 4, 5, 6 and 7 should afford sufficient guidance.

5. If a parenthetic expression is preceded by a conjunction, place the first comma before the conjunction, not after it:

    > He saw us coming, and unaware that we had learned of his treachery, greeted us with a smile.

    Place a comma before *and* or *but* introducing an independent clause:

    > The early records of the city have disappeared, and the story of its first years can no longer be reconstructed.

    > The situation is perilous, but there is still one chance of escape.

    Sentences of this type, isolated from their context, may seem to be in need of rewriting. They make complete sense when the comma is reached, and the second clause has the appearance of an afterthought. Further, *and* is the least specific of connectives. Used between independent clauses, it indicates only that a relation exists between them without defining that relation. In the example above, the relation is that of cause and effect. The two sentences might be rewritten:

    > Because the early records of the city have disappeared, the story of its first years can no longer be reconstructed.

> Although the situation is perilous, there is still one chance of escape.

Or the subordinate clauses might be replaced by phrases:

> Due to the disappearance of the early records of the city, the story of its first years can no longer be reconstructed.

> In this perilous situation, there is still one chance of escape.

But a writer may err by making his sentences too uniformly compact and periodic, and an occasional loose sentence prevents the style from becoming too formal and gives the reader a certain relief. Consequently, loose sentences of the type first quoted are common in easy, unstudied writing. But a writer should be careful not to construct too many of his sentences after this pattern.

Two-part sentences of which the second member is introduced by *as* (in the sense of because), *for*, *or*, *nor*, and *while* (in the sense of "at the same time") likewise require a comma before the conjunction.

If a dependent clause, or an introductory phrase requiring to be set off by a comma, precedes the second independent clause, no comma is needed after the conjunction:

> The situation is perilous, but if we are prepared to act promptly, there is still one chance of escape.

For two-part sentences connected by an adverb, see the next section.

6. Do not join independent clauses by a comma.

If two or more clauses, grammatically complete and not joined by a conjunction, are to form a single compound sentence, the proper mark of punctuation is a semicolon:

> Stevenson's romances are entertaining; they are full of exciting adventures.

> It is nearly half past five; we cannot reach town before dark.

It is of course equally correct to write the above as two sentences each, replacing the semicolons by periods:

> Stevenson's romances are entertaining. They are full of exciting adventures.

> It is nearly half past five. We cannot reach town before dark.

If a conjunction is inserted, the proper mark is a comma (rule 4).

> Stevenson's romances are entertaining, for they are full of exciting adventures.

> It is nearly half past five, and we cannot reach town before dark.

Note that if the second clause is preceded by an adverb, such as *accordingly*, *besides*, *so*, *then*, *therefore*, or *thus*, and not by a conjunction, the semicolon is still required.

> I had never been in the place before; so I had difficulty in finding my way about.

In general, however, it is best, in writing, to avoid using *so* in this manner; there is danger that the writer who uses it at all may use it too often. A simple correction, usually serviceable, is to omit the word *so*, and begin the first clause with *as*:

As I had never been in the place before, I had difficulty in finding my way about.

If the clauses are very short, and are alike in form, a comma is usually permissible:

Man proposes, God disposes.

The gate swung apart, the bridge fell, the portcullis was drawn up.

Do not break sentences in two. In other words, do not use periods for commas.

I met them on a Cunard liner several years ago. Coming home from Liverpool to New York.

He was an interesting talker. A man who had traveled all over the world, and lived in half a dozen countries.

In both these examples, the first period should be replaced by a comma, and the following word begun with a small letter.

It is permissible to make an emphatic word or expression serve the purpose of a sentence and to punctuate it accordingly:

Again and again he called out. No reply.

The writer must, however, be certain that the emphasis is warranted, and that he will not be suspected of a mere blunder in punctuation.

Rules 3, 4, 5, and 6 cover the most important principles in the punctuation of ordinary sentences; they should be so thoroughly mastered that their application becomes second nature.

7. A participial phrase at the beginning of a sentence must refer to the grammatical subject:

> Walking slowly down the road, he saw a woman accompanied by two children.

The word *walking* refers to the subject of the sentence, not to the woman. If the writer wishes to make it refer to the woman, he must recast the sentence:

> He saw a woman, accompanied by two children, walking slowly down the road.

Participial phrases preceded by a conjunction or by a preposition, nouns in apposition, adjectives, and adjective phrases come under the same rule if they begin the sentence.

> On arriving in Chicago, his friends met him at the station.

> When he arrived (or, On his arrival) in Chicago, his friends met him at the station.

> A soldier of proved valor, they entrusted him with the defense of the city.

> A soldier of proved valor, he was entrusted with the defense of the city.

> Young and inexperienced, the task seemed easy to me.

> Young and inexperienced, I thought the task easy.

> Without a friend to counsel him, the temptation proved irresistible.

> Without a friend to counsel him, he found the temptation irresistible.

Sentences violating this rule are often ludicrous.

Being in a dilapidated condition, I was able to buy the house very cheap.

8. Divide words at line-ends, in accordance with their formation and pronunciation.

If there is room at the end of a line for one or more syllables of a word, but not for the whole word, divide the word, unless this involves cutting off only a single letter, or cutting off only two letters of a long word. No hard and fast rule for all words can be laid down. The principles most frequently applicable are:

a. Divide the word according to its formation:

know-ledge (not knowl-edge)
Shake-speare (not Shakes-peare)
de-scribe (not des-cribe)
atmo-sphere (not atmos-phere)

b. Divide "on the vowel":

edi-ble (not ed-ible)
propo-sition
ordi-nary
espe-cial
reli-gious
oppo-nents
regu-lar
classi-fi-ca-tion (three divisions possible)
deco-rative
presi-dent

c. Divide between double letters, unless they come at the end of the simple form of the word:

Apen-nines
Cincin-nati
refer-ring
tell-ing

The treatment of consonants in combination is best shown from examples:

for-tune
pic-ture
presump-tuous
illus-tration
sub-stan-tial (either division)
indus-try
instruc-tion
sug-ges-tion
incen-diary

The student will do well to examine the syllable-division in a number of pages of any carefully printed book.

Other References

There are any number of good reference books on grammar, style and composition. If you have some doubt about whether your communications are effective, take the time to consult these references.

Some of these include:

Freeman, Lawrence H. and Bacon, Terry R. *Style Guide: Revised Edition*, Shipley Associates, Bountiful, UT 1994. ISBN 0-933427-00-X.

Hairston Maxince C., *The Scott, Foresman Handbook for Writers*, Harper Collins Publishers Inc., New York, NY, 1991. ISBN 0-673-46049-5.

Lans, Leroy L. *By All Means Communicate, 2nd Ed*. Prentice-Hall, Inc., Englewood Cliffs, NJ, 1987. ISBN 0-13-109687-7.

Partridge, Eric (New American Edition, Whitcut, Janet Ed.) *Usage and Abusage*, W.W. Norton & Co., Inc. New York, NY 1995. ISBN 0-393-03761-4

Troyka, Lynn Quitman, *Handbook for Writers*, Simon & Schuster, Prentice-Hall, Inc. Englewood Cliffs, NJ, 1987. ISBN 0-13-810409-3.

Strunk Jr., William and White, E.B., *The Elements of Style*, MacMillan Publishing Co, Inc., New York, NY, 1979. ISBN 0-02-418190-0.

BASIC PLANS AND SPECIFICATIONS

Introduction

As an elementary woodworker or shop student in junior and senior high school you may have learned that in order to construct or fabricate some project the first step was to secure a plan, or create a plan, for the intended project. From a simple bird house to a 22-story skyscraper, some form of plans and specifications must be created before the first board is cut, or the first scoop of dirt is lifted.

Plans and specifications indicate, both in general and in detail, how a project will be accomplished. Plans are generally considered to be the drawings which show the project overall and which contain specific details on how various elements of the project will interface with each other to form a completed whole. Specifications, on the other hand, are usually written text that outline what types of materials will be incorporated into the project to create the desired finished project.

A specification can be considered to be a bill of material for the elements that will be used to construct the project in accordance with the plans. In the manufacturing sector of the industry the bill of materials will list and detail all of the component parts of an assembly and are frequently contained on the drawings which detail how the component will be constructed.

To interpret these plans and specifications, you need to have an understanding of blueprints and architectural and engineering drawings. You need to know the basic symbols that are used to represent various devices, and how to measure, or scale, drawings into real life dimensions.

How Are Plans and Specifications Prepared?

At some point in time, a company or an individual, or an organization determines that they:

1. Need to build a new building (facility).

2. Need to refurbish an existing facility to accommodate increased or different usage.

When these needs become a reality, the owner (client) most often engages the services of an architect. It is the architect's responsibility to design a structure that most closely meets the client's requirements. This may include some "statement" as to the client's idea of how they wish to be perceived by their peers and their community. Let's face it, if every church building committee foresaw their responsibility as being one of creating a structure for a gathering of the faithful — churches would be square, white, simple, and maybe with a steeple.

Once an architect has been engaged, and the criteria for the project have been developed (a process which can take an inordinate amount of time), the concept of the project is developed. At this point the architect is given the responsibility for making the concept turn into reality.

Given the complexity of today's modern buildings, no single individual possesses the wherewithal to design all of the components of a building. Hence, the architect turns to various other consulting disciplines to assist in producing a set of plans and specifications that will reflect the client's expressed criteria.

Depending on the experience that the architectural firm may have "in-house" and the overall scope of the project, the architect may elect to "sub" out various subsystems designs. These might typically include:

Structural Engineer
Responsible for the mechanical structure design and the integrity of the structural members which comprise the frame of the building.

Mechanical Engineer
Responsible for the design of the HVAC (heating, ventilating and air conditioning) systems. May also have responsibility for elevators, escalators and other mechanical subsystems.

Electrical Engineer
Responsible for the design of the power distribution and other electrical subsystems within the building including lighting and various low-voltage subsystems. The electrical engineer may engage the services of an acoustical consultant, or if the building is more complex may suggest that the architect engage the services of such a consultant.

Other designers that might be included in the process would be charged with responsibility for plumbing, landscaping, acoustical analysis, fire protection systems, integrated systems design, and others.... They have interior designers, space planners, and people

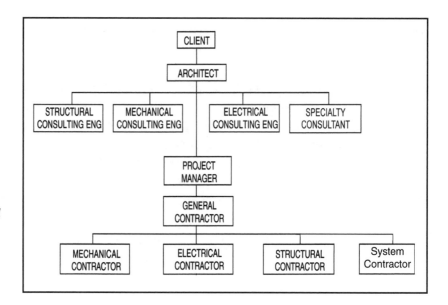

Table 1-1. *The typical team structure within the traditional bid specification marketplace.*

that specialize in kitchen design (restaurants and hospitals), transportation design, etc.

It falls on the architect to sift through all these (sometimes conflicting) design requirements and coordinate the various efforts to insure that the client gets a structure that meets the functional and artistic needs outlined in their criteria.

An Architect Does Not Have an Easy Task

Table 1-1 represents the *typical* team structure within the traditional bid specification marketplace. In real-world terms the systems contractor may find that the firm is working for the general contractor, the architect, and sometimes directly for the client.

Working from a basic set of schematics or general building outlines, the various engineering groups develop their design criteria and start to prepare drawings which represent the systems and component devices which will be incorporated into the structure to meet the criteria proposed by the client and the architect. These criteria might state, in part, for sound and communications systems:

1. "The building shall include a comprehensive paging and background music distribution system."

2. "The building shall incorporate a fire alarm and fire protection system that will be in accordance with current codes and regulations."

3. "A security system shall be provided that will control access to the building and provide surveillance for parking spaces."

4. "A comprehensive local area network (LAN) shall be included to meet existing, and foreseeable computer network interfaces."

Similar such criteria will govern the design of all of the other elements that will, in the end, be an integral part of the structure.

Ultimately, the responsibility for the final design will be assigned to some group or individual (the project manager). At that point, the designated party will have to create a set of plans and specifications that describe their portion of the project in concrete terms.

At this point the skill of the architect or project management team plays an important part in the success or failure of the project. Each of these design teams are vying for space, electrical system capacity, and superiority for access over other similar services. Electrical service cabinets should be here; fire protection stand pipes require that vertical access be provided here. Space on a ceiling is limited and must accommodate lighting fixtures, supply and return air ducts, sound system speaker components, fire alarm and fire protection components, all of which need to be installed here, and so on.

In the end, all of these diverse elements are incorporated into a single set of plans and specifications.

Plans are scaled drawings showing how all of the various elements of the building will be constructed. These incorporate all of the diverse concepts of the design team. They will generally be divided into sections which represent the contributions of the various architectural, engineering and consulting disciplines that have been involved in the design.

Divisions of the Plans and Specifications

Plans and specifications will normally be divided into sections. The major elements of a set of plans are:

A drawings. Architectural drawings of concept. Probably will include architectural details for casework, window treatments, door and hardware schedules, paint and finish schedules, carpet and wall covering schedules, etc.

S drawings. Structural drawings outlining how the structural members of the building will be erected and connected.

M drawings. Mechanical drawings detailing how the heating and air conditioning duct work and control systems shall be installed. May include elevator and escalator install information and other mechanical subsystem installation schedules.

P drawings. Plumbing drawings outlining where and how various plumbing fixtures will be installed and detailing how supply, soil, vent and sewer lines will be installed.

E drawings. Electrical drawings indicating where power, lighting, telephone, communications, sound and various electronic subsystem components will be located and connecting conduits will be installed.

Because of the sheer number of subsystems that may be included on electrical drawings, it is commonplace to separate electrical E drawings into two or more sections. One section will be used to show power distribution and lighting components and the other will show the various electrical subsystem components. If the systems are sufficiently complex, these may be further differentiated into telephone/computer, nurse call, intercommunications, television distribution, security systems, etc. Consequently, when referring to electrical drawings it may be necessary to refer to two or more sheets of drawings to define the scope of several interrelating sub-system installations.

For a major project a set of plans can be a formidable stack of drawings running literally into a hundred or more individual sheets. When tightly rolled up they can well measure 6" to 8" in diameter of 36" by 48" sheets (E-sized drawings).

Plan Scaling

Since it is impractical to show a building in full scale, some convention must be employed to represent the size and dimensions of various elements of the building. This is accomplished by "scaling." Scaling is a procedure (or method) whereby you let some defined dimension represent some other dimension. Common conventions use such scales as 1/8 inch = 1 ft., 1/16" = 1,' 1 mm = 1 m, or sometimes the drawing will be a ratio, such as 1 :1000. In the case of a small tooling drawing the reverse procedure may be employed using a scale of 1" = .001."

Within any given set of plans there may be a variation in the scales represented. A site plan might be drawn on a scale of 1/4" = 1000,' whereas an equipment cabinet detail may be represented on a scale of 1/4" = 1." Care must be taken to check that the scale being used for any particular drawing is clearly shown and understood. Some drawings which are meant to convey general information (block diagrams for instances) may be shown as N.T.S. (not to scale). Other abbreviations may be used on the plans as well. Most architects will provide a legend showing the abbreviations used on their plans. Some of the more common ones are:

    A.F.F. = Above finished floor
    B.F.C. = Below finished ceiling

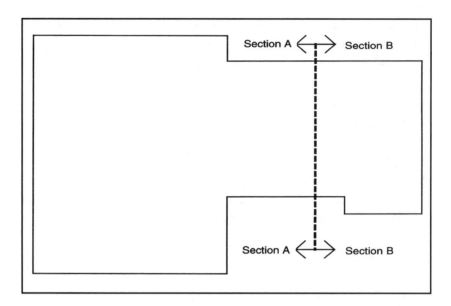

Figure 1-1. Match line representation.

N.I.C. = Not in contract
F.B.O. = Furnished by others
W.P. = Weather protected

Architectural and engineering rulers, generally referred to as scales, are available from engineering supply stores. These are scaled to allow direct footage conversion of plans with various dimensions. Anyone who has to manipulate drawing scales to real-world dimensions needs to have one of these rulers (scales) in their possession.

In some cases, it is impossible to show the entire project on one sheet of paper. In this case the architect may employ what is referred to as *match lines*. This will probably take the form of the convention as shown in *Figure 1-1*.

In other cases, the project will be divided into zones and each zone will reference to other zones in accordance with a zone plan. An example of a zone (or key plan) is shown in *Figure 1-2*.

There are some other specialized types of drawings that one will encounter in examining a set of plans. These are:

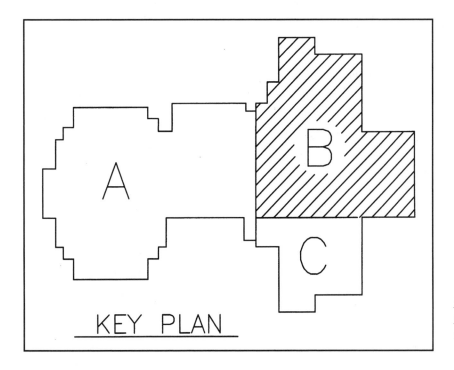

Figure 1-2.
Zone or key plan
representation.

Riser Drawings. (*Figure 1-3*) Riser drawings are single-line representatives of device layouts at various levels of the drawing. However, they can be used to show zones on a single level. Riser drawings may frequently indicate conduit sizes and cable sizes. Riser drawings should never be used to determine component counts inasmuch as they are usually representative in nature.

Reflected Ceiling Plans. (*Figure 1-4*) A reflected ceiling plan is a drawing which represents the ceiling of a structure from a position that you might view if the floor were a mirror and you were above the room looking down through the ceiling at the reflected image. Reflected ceiling plans are important in interpreting the position of various components that might need to be installed on this surface, how they relate to each other, and how the ceiling contours might dictate how such devices can be accommodated.

Symbol Representations. Inasmuch as the actual physical dimensions of each component that will be installed in the building would

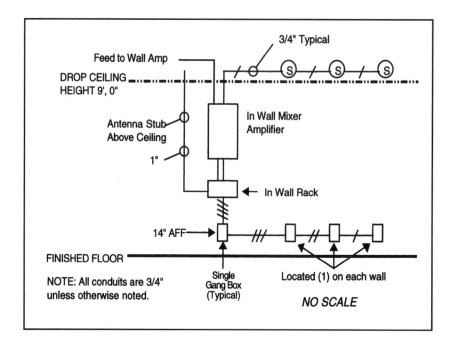

Figure 1-3. Typical
Riser diagram.

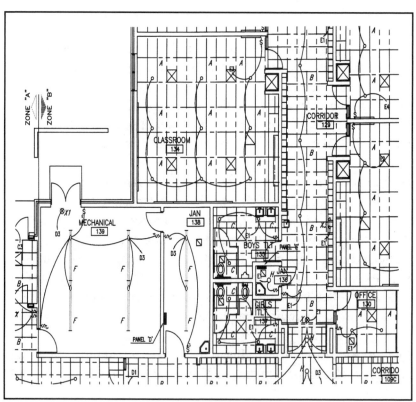

Figure 1-4. Reflected
ceiling plan.

prove much too cumbersome to try and show on a scaled drawing, a system of "icons" or representative symbols have been developed. These indicate to the experienced plan-reader what items are being used, where, and in what quantity.

A partial list of symbols for electrical devices is shown in *Table 1-2*. This list is not all inclusive and is in no sense universal. Various designers utilize their own particular set of symbols. A set of plans will include a table of symbol definitions.

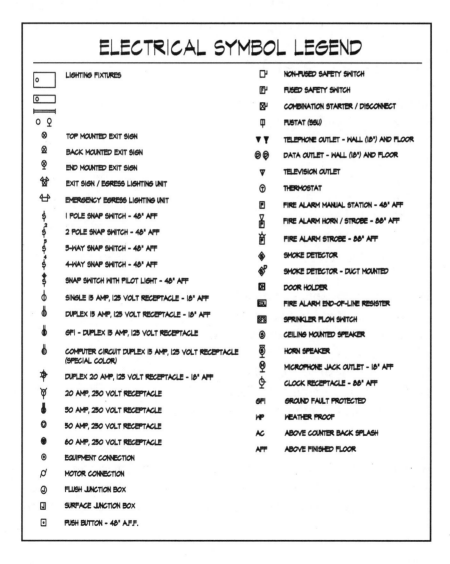

Table 1-2. *Symbols for electrical devices.*

Elements of a Specification

As previously indicated, specifications define the quality of the materials which will be employed in the particular project

Similarly, specifications are divided into sections. Each specification section covers the materials which will be used (specified to be used) for the various elements of the project. Under the Construction Specification Industries (CSI) guidelines, specifications are divided into sixteen (16) sections. These sixteen sections and the elements to which they pertain are shown below:

Specification Sections

Specification Section Covers Elements
1. General Requirements
2. Site Work
3. Concrete
4. Masonry
5. Metals
6. Wood And Plastics
7. Thermal And Moisture Protection
8. Doors And Windows
9. Finishes
10. Specialties - Toilet Compartments, Bath Accessories, Manufactured Fireplaces, etc.
11. Equipment - Fire Extinguishers, etc.
12. Furnishings
13. Special Construction - Saunas, Swimming Pools, etc.
14. Conveying Equipment - Elevators, etc.
15. Mechanical - Plumbing, HVAC, Boilers, etc.
16. Electrical

Although not officially recognized by CSI, sometimes a Section 17 is also included. Section 17 is used for "special systems" which cannot be readily accommodated within the established sixteen CSI schedules, or when it is desirable to have this scope of work performed outside of the established criteria.

Frequently special audio, video, security or control systems will be assigned to Section 17. There has been a strong emphasis within the past few years to take integrated systems out of Section 16 (electrical) and establish a distinct Section 17.

The plans and specifications for a specific project are road maps for the installers and technicians that will be working on the project. As such, they must be followed meticulously in order to avoid errors and costly rework. It might appear that the installation of a certain device might be easier, more expeditious, or more practical in a spot other than that which is shown on the plans. However, an arbitrary change could impact on some other element of the project.

Before making any changes in the field that represent deviations from the plans, the appropriate supervisor or project manager must be contacted. If the deviation is approved, an appropriate "change order" will be generated to correct the plans.

We will learn more about generating change orders and maintaining project documentation in Level II. Don't make a change unless you have supporting documentation.

Questions

1. You are asked to determine the number of ceiling loudspeakers shown on the plans for a new high school for which your firm is interested in preparing a bid.

    a. What plan sheets would you refer to for this information?
    b. What other documentation should you examine?

2. The symbol table shows a symbol for a flush wall-mount clock. None appear on the plans. What does this tell you?

3. The riser diagram for a project calls for a 1-1 1/4" conduit between two junction boxes. Upon arriving at the job site to install wire, you discover the installed conduit is only 3/4." What should you do?

4. A set of plans for the band room at the new high school shows two wall-mount loudspeakers with the note F.B.O. What does that mean?

5. The plans and specifications for a particular project appear to be in conflict with each other on some items. What would be the normal procedure for resolving the conflicts?

(Answers in Appendix A.)

Footnotes

[1] Strunk, William. 1918. *The Elements of Style.*

Chapter 2
Technical Fundamentals

- **NICET Work Element 11002, Basic Mathematics**
- **NICET's Description:** "Solve mathematical problems requiring simple addition, subtraction, multiplication, division, and raising numbers to exponential powers. Round to the correct number of significant figures, calculating percentages, reading graphs, and use simple geometric definitions and formulas. (See general mathematics textbooks.)"
- **NICET Work Element 11003, Basic Physical Science**
- **NICET's Description:** "Apply terms, definitions, and concepts for mechanics, electricity, heat, and chemistry. (Solutions may involve simple formulas found in basic physics textbooks, but will not involve algebraic manipulation or trigonometry.)"
- **NICET Work Element 11004, DC Circuits**
- **NICET's Description:** "Identify series and parallel circuits and apply Ohm's law to simple series DC circuits with linear nonreactive components. Calculate equivalent resistance of, and power consumed by, resistive devices."

Chapter Contents

DC CIRCUITS
Sources of DC Current
The Property of Resistance
Direct Current (DC) Calculations
Resistors in Series
Questions
Voltage Dividers
Questions
Resistors in Parallel
Questions
Resistors in Series/Parallel Networks
Questions
Resistance and Wire Size
Questions
Measurements in Direct-Current Circuits

BASIC MATHEMATICS

Introduction

Mathematics is indispensable to the understanding and application of the basic laws of the physical world.

There is considerable difference between understanding the mechanics of mathematics and the more difficult exercise of applying mathematics to useful applications.

It is assumed that the user of this study guide has had at least some degree of high school or college level mathematics. The material presented in the referenced text is intended to assist in the following cases:

1. Those who have an average background in high school math but have not used this discipline for several years. This material will constitute a "review" for those who fall in this category.

2. Those who have a good background in math but need a "refresher" as well as a reference manual for concepts and techniques.

3. Those who have a poor background in math who need to grasp the fundamental concepts and techniques which are applicable to real-world mathematical solutions.

If the student/applicant is not well versed in the above principles, it is suggested that he/she refer to the publication, *Basic Mathematics, with Applications to Science and Technology*. This text may be secured through McGraw-Hill Publishing. It provides the basic math essential to passing the NICET Level I examination.

Many of the basic arithmetic calculations can be performed on a four-function calculator, although a Sharp Model El-531L scientific calculator will be the calculator of choice. The student/applicant should understand the basic principles of arithmetic and the manipulation of terms. If you lack the fundamental understanding of calculations, how can you be assured that the answer on your calculator is correct? Or, worse yet, how can you correctly interpret the answer that your calculator indicates?

Simple Arithmetic

By simple arithmetic we mean adding, subtracting, multiplying and dividing whole numbers.

Competency at this level is assumed for the balance of this chapter. If you lack an understanding of these basic operations, find a tutor and bone up before proceeding.

Common Fractions

The need for greater precision in measurement led to the concept of fractions. "The thickness of a board is 3/4 inch" is a more precise statement than "the thickness of the board is between 0 and 1 inch."

In the so-called British system of measurement, the subdivisions are not multiples of 10 and such measurements are usually recorded as a common fraction. In the metric system, the subdivisions are multiples of 10 and all measurements are expressed as decimal fractions.

Fractions and Decimal Fractions

Fractions, decimal fractions — what's the difference? Basically only in the way that it is written. A common fraction, such as 1/4, has two parts; the upper part or the *numerator*, and the lower part known as the *denominator*. A fraction is really an expression of a division problem. In the fraction "1/4" the numerator is 1, in a division problem this is also known as the *dividend*; and the denominator is 4, also known as the *divisor*. If the actual division is accomplished, the result is the decimal equivalent of the fraction, or the quotient, in this case 0.25.

As you should be aware 1/4 = 0.25, and are equivalent forms of the same value.

All fractions can be expressed in a decimal form. However, the rounding that may occur could make the answer less precise. Take the example of 1/3. The decimal form of 1/3 would be 0.3333333... , the 3s repeat to infinity. Unwieldy to use in this format, we would typically round this number to 0.33 or 0.333 depending on the accuracy required.

A common fraction, such as 1/4, 1/3, and 65/66, will always be less than 1. Once a fraction becomes greater that 1 it is expressed in a form like 1 3/5; this is known as a *compound fraction*.

The compound fraction can also be expressed as an improper fraction by converting the whole digit to an equivalent fraction with a common denominator and adding the two fractions together. In the 1 3/5 example, 1 can be changed to 5/5 and added to 3/5 making a equivalent fraction of 8/5, which is the improper form of the fraction.

For a decimal equivalent, the fractional part 3/5 is solved as before and then added to the whole number. Thereby 1 3/5, becomes 1.6.

Operations With Fractions

Adding up fractions is very difficult for some people. The addition of 1/4 and 1/3 can be approached from a couple of different directions. Some people will change each fraction to its decimal equivalent and then do the addition. The simplest way, and more accurate way, is to find a *common denominator*, convert the fractions and then add the numerators.

If the denominator is the same, like 1/5 + 3/5, the result is a simple addition of the numerators, or 4/5. However, when the denominators are different, like 1/4 + 1/3, a common denominator must be found first and each fraction must be converted to its equivalent in the new denominator.

The easiest way to find a common denominator is to multiply the existing denominators together. This results in a number which is divisible evenly by both dominators, which, by the way, is the definition of a common denominator.

In the case of 1/4 + 1/3 the common denominator is 12. To convert the fractions to the new denominator, you divide the common denominator by the denominator of the known fraction and use the result to multiply the numerator. In so doing, 1/4 becomes 3/12 and 1/3 becomes 4/12. 3/12 + 4/12 = 7/12, which would be the final answer since it cannot be reduced.

Oh-oh, another new term. What is meant by reducing a fraction? Reducing a fraction to its lowest common denominator (LCD) means to find the form of the fraction when the numerator is no longer evenly divisible by any number other than 1. The fraction 25/125 can be reduced to 1/5 and the number 22/42 could be reduced to 11/21.

Finding the lowest common denominator of a fraction or a series of fractions can easily be accomplished with the aid of the calculator. The calculator (assuming the use of a Sharp EL550-L) class has a special key for entering fractions; that key is marked *ab/c*. To enter a fraction; enter the numerator first, press the *ab/c* key, then enter the denominator, then press =.

Let's try it with the fraction 25/125. Enter 25, press *ab/c*, then enter 125, then =. The calculator should be showing your entry on the top line [25r125 =] and the LCD form of [1r5] on the second line. The r character shown is equivalent to a "/" or a division sign.

To add two fractions enter the first numerator and press the *ab/c* key. Then enter the denominator and press the + sign. Enter the second fraction just like the first, then press =.

For our earlier problem of 1/4 + 1/3, you should get the answer of [7r12] or 7/12. Wow, the calculator did the work of finding the common dominator along with solving the problem. In fact, the calculator will always give you the answer in the lowest common denominator form.

The same process works with all standard operations; addition, subtraction, multiplication and division.

Questions

1. Find the answers to the following problems using the LCD.

    a.   1/4 +2/3 - 1/16
    b.   1 1/5 + 5/10 + 7/3
    c.   3/5 x 7/8
    d.   5/7 ÷ 1/3

2. Give the answer to these problems as improper fractions.

    a.   (2/5 + 4/7) ÷ 20/22
    b.   23/6 x 4/9 ÷ 1/3
    c.   231/125 - 15/20
    d.   6/9 + 7/8 + 5/4 - 1/3

3. Find the product of the numbers in each of the following sets to four rounded decimal places:

    a.   3.567, 45.65
    b.   1.498, 0.05467
    c.   56.2, 0.015

4. Divide the first of each of the following pairs of numbers by the second to four rounded decimal places:

    a. 4.567, 34.46
    b. 4.006, 0.0063
    c. 70,1.609

5. Round off the following to the specified number of places:

    a. 54.109 to two places
    b. 762.50 to an integer
    c. 0.287750 to three places

6. Reduce the following fractions to their lowest terms:

    a. 36/27
    b. 60/180
    c. 128/224

7. Given the fractions 2/5, 1/4, 3/8:

    a. Determine their sum.
    b. Determine their product.
    c. Subtract the third from the sum of the first two.
    d. Find the sum of the first and third, then divide this sum by the second.

(Answers in Appendix A.)

Percentages

Percent (%) is a special type of a fraction with 100 as a denominator. Thus, 20 percent (20%) means 20 hundredths or 20/100. A fractional measure can always be expressed as a percent by writing an equivalent fraction with 100 as its denominator:

$$3/5 = (3 \times 20)/(5 \times 20) = 60/100 = 60\%$$

Since a decimal may be written as a common fraction, it is also possible to express a decimal fraction as an equivalent percent.

$$0.24 = 24/100 = 24\%; \; 1.15 = 115/100 = 115\%$$

Percent is frequently used to express an estimate of accuracy between two determinations of an unknown quantity.

Example: A resistor is measured by the voltmeter-ammeter method and by using a Wheatstone bridge:

$$R_{VA} = 22.4 \text{ ohms}; \; R_W = 22.2 \text{ ohms}$$

What is the percentage difference between the two methods? The data does not tell us which method is more accurate, hence we use the average as base.

$$R_{average} = (22.4+22.2)/2 = 22.3 \text{ ohms}$$

$$\% \text{ difference} = \frac{22.4 - 22.2}{22.3} \times 100 = 0.897 = 0.9\%$$

Questions

1. A wattmeter has an accuracy of ±2% of the full-scale reading. The scale range is 0 - 500 watts. If a measurement of 150 watts is made with this instrument, within what limits does the true value lie? ~~answer~~

2. An electrical resistance decreased 2%. If the present value of the resistance is 74.8 ohms, what was the value before the change took place?

3. A sample of water is decomposed into 0.965g of oxygen and 0.120g of hydrogen. Calculate the percent of oxygen and hydrogen in the sample of water.

(Answers in Appendix A.)

Measurements

There are many different devices for taking measurements of physical quantities: rulers, balances, voltmeters, thermometers, barometers, etc. Instruments of this type have scales graduated in some form of standard unit(s).

Mensuration is the process of making a measurement.

Mensuration by means of an instrument necessitates an interpolation of the distance between the initial and final position of the needle or the barometric level of a column of mercury.

Exactness in mensuration is a relative term; even the most careful observer using the most delicate of instruments cannot make a perfect measurement of a continuous quantity. Thus, the record of every measurement must consist of a number, an estimate of its uncertainty, and a unit.

Every measurement of a continuous quantity is only an approximation of the true or absolute value of that measure. The inherent uncertainties in mensuration are called experimental errors or uncertainties. These can be either systematic or random.

Accuracy and precision can be expressed as an "absolute error" and/or "relative error." *Absolute error* is the difference between an experimental value of a physical quantity and the accepted value. The *relative error* or *relative uncertainty* of an experimental value is the ratio of the absolute error to the accepted value.

Let's use the following questions to test our understanding of the previous concept.

Questions

1. The speed of sound in air at 25° C has a "handbook" value of 346 m/sec. A laboratory measurement at the same temperature

results in a value of 342 m/sec. Calculate (a) the absolute error; (b) the relative error.

2. A voltmeter has two scales: 0 to 5 and 0 to 50. The reading uncertainty of the 0 to 5 volt scale is 0.05 volt; the uncertainty of the 0 to 50 volt scale is 0.2 volt. If a reading of 2.5 volt is taken with each of the two scales, what are the two relative uncertainties?

(Answers in Appendix A.)

Exponents and Radicals

Exponential notation may be used to show that the same number is repeated several times in a product. Thus 5 x 5 is simplified to 5^2 and r x r x r = r^3.

The positive integer which shows the number of times the base is used as a factor is called the positive integral exponent. In 5^2 the exponent is 2; in r^3 it is 3. In the product a^n, the factor "a" occurs "n" times.

The notation 5^2 is read "the second power of 5" or "5 raised to the second power" or "5 squared". The number r^3 is read "the third power of r" or "r raised to the third power" or "r cubed." The product a^n is read "a raised to the nth power" or the "n^{th} power of a."

By definition:

$a^0 = 1$, if a is not zero.

$a^{-n} = 1/a^n$, if n is a positive integer and a is not zero

If two numbers a and b satisfy the equation $a^n = b$ with n as a positive integer, then a is defined as an nth root of b.

If $a^n = b$ then $a = \sqrt[n]{b}$

For a value where the exponent of the number is -1, like 5^{-1}, the result is the reciprocal of the number or the same as 1/5, or 0.2. The calculator uses this notation for finding a reciprocal, instead of the *1/x* key found on some other calculators.

The method of finding the reciprocal of the number 200 would be to enter the number 200, then press the *2nd F* key, then the X^2 key, and then =. The second function of the X^2 key is the X^{-1} function. Your display should show [200^{-1} =] on the top line and the answer 0.005 on the bottom line.

Example:

$^2\sqrt{9} = 3$ also $2^{3/2} = 2\sqrt{2^3} = {}^2\sqrt{8} = 2.83$
$^2\sqrt{9}$ is read as the square root of 9
$^2\sqrt{2^3}$ is read as the square root of 2 cubed

In the equation $2\sqrt{9}$, the 2 is the radical, which denotes the root to be found. In normal notation the root of 2, or square root, is understood and not displayed. Any other root, or radical, must be displayed for the proper operation to take place.

There is a long and involved arithmetical method of determining the square root or the nth root of any number. However, with the advent of modern scientific calculators, the solution to these types of exercises can be determined rapidly.

On your calculator, to square any number or raise a number to the power of 2, enter the number then press the X^2 key, then =.

To raise a number to any number other than 2, you will use the Y^X key. In an example of raising the number 3 to the 5th power, enter the number 3, then press the Y^X key, then enter 5 and press =. The result 243 will be displayed along with your entry in the format [3^5 =].

To enter a negative exponent, press the **+/-** key after you enter the exponent, but before you touch the **=** key.

To find the square root of 25 on your calculator, press the √ key first then enter 25, then press **=**. Your display should show [√25 =] on the first line and the answer of 5 on the second. The square root of any number may be found this way.

The cubed root of a number is found in the same way, only you need to use the *2nd F* key to shift to the second function of 3√ using the same √ key. So, to find the cubed root of 216, press *2nd F*, then the √ key, then enter 216 and press **=**. Your display should show [3√216=] on the first line and the answer 6 on the second.

Finally, for finding any other root of a number you will use the *2nd F* key with the *Yˣ* key to get the ˣ√ function.

Questions

1. Write each of the following in exponential form:

   a. 3 x 3 x 3 x 3 x 3

   b. $\dfrac{1}{5 \times 5 \times 5 \times 5 \times 5}$

   c. $\dfrac{1}{16 \times 16 \times 16}$

2. Write the results of the indicated operation in simplified exponential form:

   a. 2^5 x 2^{-8} x 2^6

   b. 3^{-5} x 3^6 x 3^{-2}

   c. $(4p)(4p)^3(4p)^{-4}$

(Answers in Appendix A.)

Scientific Notation

A number is said to be in *scientific notation* when it is expressed as a power of ten. There are several advantages of expressing numbers in this notation.

1. Very large and very small numbers can be written more compactly and the arithmetical operations are greatly simplified.

2. An estimation of the result of an involved computation can be obtained quickly.

3. Scientific notation is used widely to indicate the precision of measurements.

Millions	1,000,000	1×10^6
Hundred Thousands	100,000	1×10^5
Ten Thousands	10,000	1×10^4
Thousands	1,000	1×10^3
Hundreds	100	1×10^2
Tens	10	1×10^1
Units	1	
Decimal point	0.0	
Tenths	0.1	1×10^{-1}
Hundredths	0.01	1×10^{-2}
Thousandths	0.001	1×10^{-3}
Ten-Thousandths	0.0001	1×10^{-4}
Hundred-Thousandths	0.00001	1×10^{-5}
Millionths	0.000001	1×10^{-6}

For a number to be correctly written in scientific notation, the accepted practice is for the *mantissa* to be greater than, or equal to 1, but less than 10. Mantissa? What's that?

A number like 840,000,000 is expressed in scientific notation as 8.4×10^8. The 8.4, or part of the expression before the x sign, is the mantissa. Although you may sometimes see this number written as 84 x 10^7, this is not the most correct way of doing it in scientific notation.

Numbers may be directly entered into the calculator in scientific notation. In addition, the calculator may be set to display all answers in scientific notation.

Check your calculator's users guide for the method of changing your display to scientific notation.

To enter a number like 8.4 x 10^8 into your calculator; first enter the 8.4, then press the *Exp* key, and then enter the exponent 8. When you pressed the *Exp* key you should see the symbol [x10^{00}] show up behind the 8.4. Your next keystrokes are used as the number for the exponent to the 10. If the exponent is to be a negative number, press the **+/-** key after entering the exponent.

Like fractions, numbers in scientific notation may be entered in the manner indicated and then operations, such as addition, subtraction, multiplication and division, may be accomplished. Answers will be given in scientific notation only if you have changed the calculator display mode. Otherwise, answers will be in normal floating decimal point format.

Question

1. The electric energy generated in 1960 in the United States was 842 billion kilowatt-hours.

    a. Express this quantity in scientific notation for kilowatt-hours.
    b. The United States population in 1960 was 179.3 million. Express this quantity in scientific notation.
    c. Calculate the electrical energy generated per capita in the United States in 1960. Express this quantity in scientific notation.

(Answers in Appendix A.)

Ratio and Proportion

A *ratio* is the comparison by division of two quantities expressed in the same unit.

The idea of ratio is also used in connection with maps and drawings. If a map indicates a scale of 1 in. = 10 mi, the ratio of distances on the map to actual ground distances is 1 to 633,600 (10 miles = 633,600 inches). Both quantities in a ratio must have the same unit of measure.

The symbolism for the ratio of a number *a* to a number *b* is *a:b* or *a/b*. Ratios may be manipulated in the same fashion as common fractions or decimal equivalents.

A statement that two ratios are equal is called a proportion. For example the ratios 2:3 and 4:6 are equal and form a proportion, this may be written as:

2:3 = 4:6 or 2:3::4:6 or 2/3 = 4/6

The proportion a:b = c:d is read "a is to b as c is to d." (b and c) are called the *means* and (a and d) are called the *extremes*.

The product of the means is equal to the product of the extremes. Hence if we are given any three members of a proportion, we can determine the fourth member.

Example: determine b if 3:5 = b:15; 3/5 = b/15:

3 = b
5 = 15
Cross multiplying (3 x 15) = 45 and (b x 5) = b5
5b = 45
Dividing both sides by 5 results in b = 9.

When a quantity is changed in some ratio and another quantity is changed in the same ratio then the two quantities are said to be in direct proportion to each other. If the diameter of a circle is doubled, the circumference will also double. Hence, the circumference is directly proportional to the diameter, and the circumference varies directly as to the diameter.

When a given quantity is changed in some ratio and another quantity is changed in the inverse ratio, then the two quantities are said to be in inverse proportion to each other.

One of the elementary laws of acoustics is that as the distance from a sound source is doubled then the sound pressure level is halved. In the sound and communication industry, the inverse square law of acoustics is one of the most frequently used examples of inverse proportion.

$$\frac{\text{Distance}_1}{\text{Distance}_2} = \frac{\text{SPL}_2}{\text{SPL}_1}$$

Questions

1. Solve the following sets of proportions for the unknown member:

    a. x:5 :: 6:15
    b. 5/m = 1/72
    c. 4:90 = t:130

2. If 25 feet of wire weighs 2-3/4 pounds, what is the weight of 6 feet 4 inches of this wire?

3. A stake 10 feet high casts a shadow 8 feet long at the same time that a tree casts a shadow 60 feet long. What is the height of the tree?

4. The current in an electrical circuit is in inverse ratio to the resistance. If the current is 2.4 amperes when the resistance is 45 ohms, what is the current when the resistance is 25 ohms?

5. If the scale of a map is 1 cm = 15 km, determine the distance on the map corresponding to an actual distance of 135 km.

6. The rotational speeds of two meshed gears vary inversely as to their number of teeth. If a gear with 24 teeth rotates at 420 rpm, how many teeth should a meshing gear have to rotate at 360 rpm?

(Answers in Appendix A.)

Area and Volume

The *area* of a simple two-dimensional geometric shape is usually the product of the length and width of the shape.

For instance, a square with sides of 5 feet will have a total area of 5' x 5' or 25 sq. ft. A rectangle that is 5 yards tall and 8 yard wide, will have an area of 5 yd. x 8 yd. or 40 sq. yards. Note that the answers are always given in square units and the numbers being multiplied must be in the same units.

The area of a triangle is calculated differently. The formula usually used is 1/2BH, or 1/2 the base of the triangle times its height. So in a triangle with a base dimension of 8 inches and a height of 12 inches, the area would be 1/2(8)(12) or 4 x 12 = 48 sq. inches.

The area of a circle is found by the formula πr^2 or pi times the radius of the circle squared. *Pi* is a constant approximately equal to 22/7 or 3.1416. You will see this constant used all the time in determining areas of circles and volumes of spheres. The radius of a circle is the length of a ray extended from the center of the circle to a point on its circumference. The radius will be half the value of the diameter of a circle, which is a line that bisects the circle.

So, for a circle with the diameter of 10 meters, the area would be $\pi(5^2)$ or π times 25 = 78.54 sq. meters.

Area formulas for other shapes can be found in the Level II math section.

The *volume* of an object is the total space inside a three-dimensional object. Like area we use the dimension of length and width and add the dimension of depth. The volume of a rectangular object is found by multiplying the length times the width times the depth; or by using the area calculation already found for the length times the width, then multiplying by the depth.

In a rectangle that is 5 yards tall and 8 yards wide and with a depth of 6 yards, we would calculate 5 x 8 = 40 sq. yards times 6 yards to get a total volume of 240 cubic yards, or 240^3 yards. Note the answer is in cubic units or units cubed.

Volumes for prisms are found by finding the area of the side triangle and multiplying it by its depth. Using the triangle we mentioned before with the base of 8" and a height of 12", if we add the depth dimension of 15", we can find the volume by multiplying the area of 48 sq. in. by the depth of 15" and getting a volume of 720 cubic inches or 720^3 in.

Again, additional formulas for finding volumes of spheres, cones, etc., may be found in the math chapter in the Level II handbook. You should not need those formulas to answer Level I questions if taking the NICET exam.

Questions

1. A rectangular plate has the dimensions of 4.5" x 18.35."
    Calculate the area of the plate in scientific notation to three significant digits.

2. Express this area in scientific notation in cm^2 and m^2. (1" = 2.540 cm and 1 m =100 cm.)

(Answers in Appendix A.)

BASIC PHYSICAL SCIENCE

Introduction

Physics is at the base of all other physical sciences and much of technology because it deals both with things that may be treated as fundamental in a specific problem and with the laws than govern their interactions, individually and in combination.

To some people the discipline of physical science is considered akin to mathematics or in some way a "higher" degree of mathematics. Certainly physics relies on mathematics to solve physical science problems, and the theories of physics are frequently expressed in mathematical expressions. However, the use of mathematical equations are employed to express known relationships between real life objects and the forces and conditions which work upon and between them.

Physics is, in many ways, more "practical" than pure mathematics and relies more on conceptualization of processes. Until you can understand the concepts and/or verify the theories by experimentation, any mathematical computations that you perform are virtually useless. Obviously, if you have no appreciation for the interrelationship of electrical energy, current and resistance, any calculations you might perform might be interesting mathematical exercises but the results are not going to be very meaningful.

The scope of classical physics is broken down into a few broad categories, such as mechanics, thermodynamics, electricity and magnetism, and optics. Modern physics expands this range to include relativistic mechanics, atomic physics and quantum mechanics, and nuclear and particle physics. In this text we will be dealing with classical physics only.

The Physical Quantities

The building blocks of physics are the physical quantities that we use to express the laws of physics.

Force, *work*, and *energy* are general physical quantities.

The essential quantities of the mechanical group are *time*, *length*, *area*, *volume*, *mass*, *density*, *speed*, *acceleration*, and *power*.

These quantities fall into two distinct classes:

1. Those that are fundamental. The units of time, units of length, and units of mass are fundamental. These units are defined against an existing standard.

2. Those that are defined as a function of their relationship with the fundamental.

Time
The unit of time has been defined as the interval (on earth) that comprises a solar day (one revolution of the earth on its axis). This measured quantity is then further defined in terms of hours, minutes and seconds (and in some cases fractional expressions of seconds). A (mean solar) second is equal to 1/86,400 of a (mean solar) day. Time defined in terms of the rotation of the earth is called *universal time* (UT).

In 1967, the *second*, based on the cesium clock, which uses the periodic atomic vibration of an atom of cesium 133, was adopted as an international standard by the 13th General Conference on Weights and Measures. This standard allows precision that would vary no more than 1 second in 6000 years.

Length
Likewise, length is an arbitrary, human-introduced concept of distance. The Imperial system (English system) grew out of early

medieval England and includes such terms as *rod, stone, bushel, chain, slug,* etc., terms that are somewhat of a curiosity today and which are not in general use. Unless you're a horse race fan the concept of a furlong is probably not ingrained into your memory bank.

The more general use today is to express length and the derivatives of volume and area in terms of the SI units which comes from Systeme International d'Unites — The International System of Units. The SI is based on the metric system and hence has a relationship to the base 10.

The SI unit of length is the *meter*. Historically, the meter was intended to be one ten-millionth of the distance from the north pole to the equator along the meridian line through Paris. However, accurate measurements show the standard meter bar differs slightly (about 0.23%) from this value. The first international standard of length was a bar of a platinum-iridium alloy called the *standard meter*, and it was kept at the International Bureau of Weights and Measures. This standard has been replaced with a more accurate standard using the wavelength, in vacuum, of a particular radiation emitted by atoms of a particular isotope of krypton in electrical discharge. Specifically, one meter is now defined to be 1,650763.73 wavelengths of this light.

From the unit of length we also derive mathematical units for area or volume.

For area, such units are the square inch (sq. in., in^2); square meter (-m^2).

For volume, such units are the cubic foot (cu. ft., ft^3); the cubic meter (cu. m, m^3); gallon and liter.

Note the terms *cubic meter, cubic feet,* etc., are usually used for dry measure — a cubic yard of cement, a block of stone measuring 2

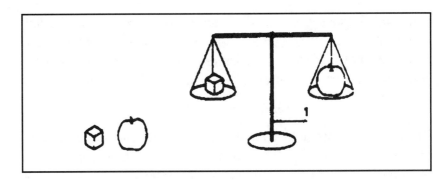

Figure 2-1.
An equal arm balance.

cubic meters, etc., whereas the terms *liter, gallon, quarts,* and *milliliters* usually denote fluid volumes.

Mass
The concept of the third fundamental unit, the unit of mass, is frequently confused with the concept of force, and the everyday term *weight* may be mistakenly used to denote mass.

Mass is defined as the amount of matter an object contains. There is a cylinder of platinum-iridium preserved in the International Bureau of Weights and Measures in Sevres, France, known as the "Kilogram of the Archive." This is our standard unit of mass. This standard is 1 kg or 1,000 grams.

Secondary standards are sent to standardizing laboratories in other countries and the masses of other bodies can be found by an equal-arm balance technique to a precision of two parts in 10^8. See *Figure 2-1*.

All other units or quantities are derivations of these three fundamental units.

Weight, Density and Specific Gravity
The earth exerts a force of attraction (*gravity*) on the mass of all bodies. In practice we frequently compare these forces as the weights of the objects rather than making a direct comparison of the masses of the objects. An object in space is weightless; however, its mass

in space is the same as its mass on earth. The greater its mass, the greater the force required to alter its speed of movement.

On earth, weight is the measure of the force with which gravity pulls an object toward the earth's center. Gravitational force decreases with distance from the earth's center, so the further away an object moves from the earth's center the lighter the object becomes. Weight can be measured with a spring balance which is calibrated using a known mass at sea level. See *Figure 2-2*.

Gravity conforms to the definition of *force* as being an entity that can alter an object's size, shape, speed, or direction.

The international unit of force is the *Newton* (N). One N is the amount of force needed to give a mass of one kilogram an acceleration of one meter per second per second (1 m/sec^2), i.e., each second the mass travels 1 m/sec faster than during the previous second.

The concept of gravity is attributed to Sir Isaac Newton (English scientist and mathematician 1642-1727). His theory of gravitation was supposedly inspired by his observation of an apple falling from a tree in his garden one afternoon while he was having tea.

Using the apple idea, why does the apple fall? Or more specifically, why does the apple fall downward? As noted above, an apple

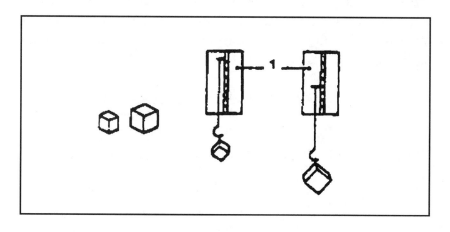

Figure 2-2. *A spring balance.*

has a much smaller mass than the earth; hence, the earth's gravitational force attracts the apple towards the center of the earth. Earth's gravity at the surface of earth exerts a pull of ≈9.8 Newtons on a mass of 1 kg.

The earth's mass is 6 million billion billion kilograms (6 x 10²⁴). If the mass of the apple is 0.1 kg then the force (pull) of earth's gravity gives the apple its weight. In scientific terms the weight of an object is measured in Newtons. The weight of an object is therefore the gravitational force on the object and is equal to the object's mass multiplied by the gravitational field strength.

Gravity ≈ 10 m/s²
The weight of the apple = mass x gravity
The weight of the apple »1 N

Density is defined as the mass per unit volume.

Density = *mass*
             volume

In practice we seldom use density in the absolute sense but instead use the term specific gravity. The *specific gravity* of any material is the ratio of the weight of a given volume of that material to the weight of an equal volume of water. For example, the specific gravity of a cubic foot of cast iron which weighs approximately 500 lbs. has a specific gravity of 7.7, inasmuch as a cubic foot of water weights approximately 62.5 lbs. Specific gravity is therefore simply the ratio of the weight of one property (in similar units) to the weight of a given standard.

Specific Gravity = Weight of a volume of a substance
                          Weight of an equal volume of water

1 ft³ of water ≈62.5 lbs and 1 ft³ of water ≈0.028 m³ of water which is ≈28.375 kg.

Mechanics

The oldest of the physical sciences is mechanics. It is the study of the motion of objects. When we describe motion we are dealing with the section of mechanics called *kinematics*. Another part of mechanics is "dynamics," which deals with the forces associated with motion and the study of the properties of the moving object. Also a part of mechanics is a division called "statics," which deals with the effects of stresses and strains on matter at rest.

In the study of kinematics and dynamics we will be dealing with Sir Isaac Newton's three "Laws of Motion." Also known as the "Laws of Inertia."

Simply stated, the laws are:

1. Objects in motion (or at rest) remain in motion (or at rest) unless acted upon by an external force.

2. Any external force F (commonly designated by a bold-type uppercase letter F) will produce an acceleration a (commonly designated by a bold-type lowercase letter a) of a body of mass m according to: $F = ma$.

3. Every action is matched by an equal and opposite reaction.

Force
Force is that influence which tends to set any body in motion, or that changes the direction and/or speed of any body already in motion. That mass of platinum-iridium referred to earlier as the standard unit of mass is drawn toward the center of the earth by the force of gravity. If it were free to fall, the force of gravity would give it an acceleration of 9.8m per second per second. That is, at the end of the first second from the time it started to fall it would attain a velocity of 9.8 meters per second (32 feet per second) and at the end of the next second it would have a velocity of 19.6 meters per second (64 feet per second).

When the forces acting upon an object in any one direction are equal to those acting in the opposite direction, the object is at rest or is in *equilibrium*. When these forces become unbalanced, the object is set in motion and work is performed.

The SI unit of force, in the MKS system, is the Newton (N). One N is the amount of force needed to give a mass of one kilogram an acceleration of one meter per second per second. In the CGS measurement standard, the unit of force is the *dyne*, where 1 dyne is the amount of force needed to move one gram an acceleration of one centimeter per second per second. By comparison, $1N = 10^5$ dynes.

Work
Work is done when a force moves a body in the direction of the force. (*Figure 2-3*)

Work is measured by the product of the force and the distance through which it acts:

work = (force)(displacement) or W = Fd

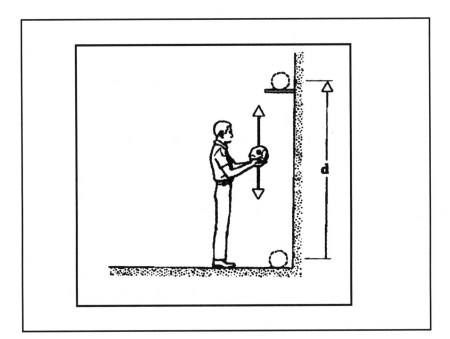

Figure 2-3. Work is done when a force moves a body in the direction of the force.

If a six-pound hammer is raised from the floor to a bench three feet in height, six times three, (6 x 3), or eighteen foot-pounds of work has been performed.

If the vertical distance between two floors of a building is 12 feet and a man weighing 150 pounds ascends from one floor to another he performs (12 x 150) or 1800 foot-pounds of work.

Note: The horizontal distance which he might traverse in climbing the stairs is not a factor of the equation. The distance must be calculated or measured parallel to the direction of the force.

Work in its scientific sense is measured by the result and not by the effort or the exertion. Someone might attempt to move a weight until fatigued; but no mechanical work has been performed unless the effort to overcome the effect of gravity, or the effect of inertia and friction which tends to hold the object at rest, or any other force that may be acting upon the object has been overcome.

The SI unit for work is the joule (J), which equals one newton-meter (Nm).

Power
Power is the time rate of doing work.

Using the SI System:

$$\text{Power(watts)} = \frac{\text{Work (joules)}}{\text{Time (secs)}} \text{ or } P = \frac{W}{t}$$

One watt equals 1 joule per second or 1 J/s

The English unit of power, therefore, not only requires the fundamental work units of foot and pound, but must also include time.

If we were to say that a piece of machinery is capable of doing 33,000 foot-pounds of work without stating a dimension of time, then we have no conception of how powerful the machine might be. It might be a very small machine that takes 15 days to do 33,000 foot-pounds of work, or it might be a very powerful machine that can perform this work in a matter of seconds.

By definition, if a machine can perform 33,000 foot-pounds of work in one minute, or 550 ft. lbs./s, it is rated at one horsepower. One wonders at how, or when, someone observed that a horse could move 16.5 tons a distance of one foot in one minute; but obviously someone, at some time, with some horse, hit on this concept.

One horsepower is equal to approximately 746 W or 3/4 of a kilo-watt.

Work can also be expressed in units of (power x time). For example, the term *kilowatt-hour* is an expression of work done in 1 hour by an agent working at a constant rate of 1 kilowatt.

Questions

1. On the way to a job, Charlie drags a box full of cable at a constant speed, 15.0m from the shop to a service van with a horizontal force of 95 N. How much work does Charlie perform to overcome the force of friction?

2. Joe is a big man having a mass of 130 kg. He is asked to go up two floors of a building. Each flight of stairs is 5.0 m. How much work does Joe perform in ascending the stairs? (hint: how much does Joe weigh?)

3. Atlas and Hercules are two carnival sideshow strong men. Each lifts 200 kg barbells 2.0 m off the ground. Atlas lifts his weights in 1.00 second and Hercules lifts his in 3.0 seconds:

   a. Which strong man does more work?
   b. Which of the two is more powerful?

(Answers in Appendix A.)

Acceleration

In the same sense that power is the time rate of doing work, speed is the time rate of traversing a distance.

If a moving object, a car or a train for example, should either acquire more speed or slow down, it would be accelerated.

Acceleration is defined as the time rate at which an object changes speed. The car or train in the above example is positively accelerated when getting up to speed, and is negatively accelerated when slowing down. So even if we refer to an automobile's speed-up mechanism as an accelerator and its slow-down mechanism as a brake, by definition in physical science the vehicle itself is still being either positively or negatively accelerated.

The quantity of acceleration is an ideal example of a measurement that requires more than a mere numerical comparison. If in one minute of time the vehicle increases its speed one mile per minute, it would have a positive acceleration of plus one mile - per minute - per minute (1 mi/min^2).

$$\text{Average acceleration} = \frac{\text{change in velocity}}{\text{elapsed time}} \text{ or } a = \frac{\Delta v}{\Delta t} = \frac{v_f - v_o}{t_f - t_o}$$

— where v_f means final velocity and v_o means initial velocity.

Question

1. You're driving down the street at a speed (velocity) of 30 m/s when a ball rolls out into the street in front of you. You slam on the brakes and come to a stop in 3.0 s. At what rate of acceleration did you come to a stop?

(Answers in Appendix A.)

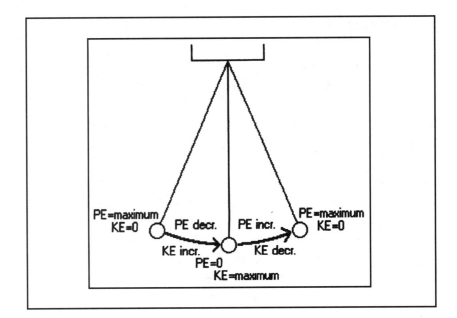

Figure 2-4.
Relationship between
potential energy (PE)
and kinetic energy
(KE) for a swinging
pendulum.

Energy

Energy is the ability to do work or is stored work in some form and may therefore be expressed in the same units as work. The form that bodies in motion possess is called kinetic energy. Energy may be stored in the form of *potential* energy, as it is in a compressed spring (see *Figure 2-4*). Chemical systems possess *internal* energy, which can be converted by various devices into useful work; for example, a fuel such as gasoline can be burned in an engine to propel a vehicle. Heat energy may be absorbed or released when the internal energy of a system changes while work is done on or by the system.

Kinetic Energy
Masses in motion possess kinetic energy. For example, an object of mass m moving with the speed v possesses kinetic energy one-half mv squared. ($KE = 1/2mv^2$)

Potential Energy
A suspended weight or a compressed spring are said to have stored,

or potential, mechanical energy by virtue of their ability to at some point in time perform useful work.

Internal Energy
Energy contained in a system by virtue of the motions of, and forces between, the individual atoms and molecules of the system is called internal energy.

Types of Energy
Energy can also exist in forms other than mechanical. There are six forms of energy, these are illustrated along with examples of each and the interrelations between them in the following chart.

The following table lists the six types of energy and examples of each, and *Figure 2-5* illustrates the relationships between these forms of energy:

Energy Type	Example
1. *Mechanical energy-kinetic or potential energy.*	*Dynamo* *Turbine*
2. *Electrical energy: the energy produced by electrons moving from atom to atom.*	*Motor* *Electric heater* *Light bulb*
3. *Nuclear energy: the energy stored in the nucleus of an atom.*	*Power Station* *Bomb*
4. *Heat energy: the energy produced by the random movement of a substance's atoms. The faster they move, the hotter they become.*	*Anything red or white hot.*
5. *Radiant energy: consists of rays, waves or particles, especially forms of electromagnetic radiation, such*	*Solar Panels* *Solar Cells* *Photosynthesis*

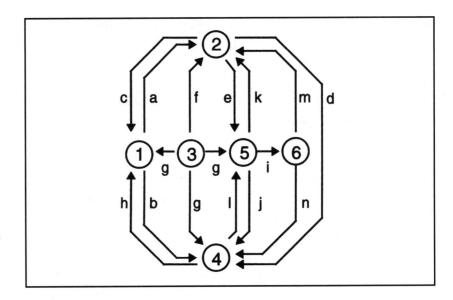

Figure 2-5. Six types of energy and their relationships. See list of energy types for assignments.

as infrared radiation, light, ultraviolet radiation, X-rays, gamma rays, and cosmic rays.

6. *Chemical energy: the energy stored in*　　*Batteries*
   an atom or molecule and released by　　*Gas or Oil Furnace*
   a chemical reaction.

Thermodynamics

Thermodynamics is the branch of science concerned with the nature of heat and its conversion into other forms of energy. Heat is a form of energy associated with the positions and motion of the molecules of a body. The total energy that a body contains as a result of the positions and the motions of its molecules is called its internal energy. The first law of thermodynamics states that in any process the change in a system's internal energy is equal to the heat absorbed from the environment minus the work done on the environment. This law is a general form of the law of conservation of energy.

The relationship between heat and work was established when it was noticed that when a fixed amount of work is done on a fluid —

stirring it, for example — a fixed amount of heat is always generated. The relationship between heat energy, which is usually measured by observing temperature changes in an object, and mechanical work is called the mechanical equivalent of heat. The relationship is expressed using the term *calorie*. A calorie is the heat required to raise the temperature of 1 gram of water 1 degree C.

1 calorie = 4.186 joules

The British thermal unit, abbreviated BTU, is the unit of measurement for the quantity of heat in the customary system of Imperial (English) units of measurement.

A BTU used to be defined as that quantity of heat required to raise the temperature of one pound of water by 1 degree Fahrenheit.

The BTU was refined in 1929, in terms of electrical units. The BTU is equivalent to 251.996 calories, 778.26 pounds, or .2928 watt-hours.

A pound (0.454 kilograms) of good coal when burned should yield 14,000 to 15,000 BTU; a pound of gasoline or other fuel oil will yield approximately 19,000 BTU.

You might well ask, why is this important in the understanding of sound and communications systems? Electronic equipment does generate heat. Some components such as power amplifiers generate a considerable amount of heat. The calculation of heat buildup has considerable bearing on how the equipment must be spaced within a rack or enclosure. In larger installations, the equipment rack(s') heat buildup (in BTUs) can have an effect on the cooling and ventilation of the equipment room.

Efficiency
Energy may be converted from one form to another but can never be produced or destroyed.

The best known devices available for converting mechanical, chemical, electrical, or any other form of energy into another form of energy are, to varying degrees, rather inefficient devices. Most of the energy is transformed into thermal, or heat, energy rather than into the form of useful work. Hence, efficiency is a measurement of the power in, versus the power out.

$$\%Eff = \frac{Power\ Out}{Power\ In} \times 100$$

Let's take a look at a good example in our own field of endeavor, which graphically illustrates this concept. It has become all the rage to rate loudspeakers in various ways to denote "power handling capacity." This is probably the most useless specification parameter for a loudspeaker. The purpose of a loudspeaker is to produce acoustical power that our ears translate into sound. The input power to the loudspeaker is electrical, having been amplified by some form of electrical power. Hence, a loudspeaker is a transducer or converter which takes electrical power, transforms it into mechanical power that causes a movement of the loudspeaker cone, which in turn translates the cone movement into radiant power that becomes what we perceive as acoustical power.

The best low-frequency (woofer) loudspeakers available have an efficiency in the neighborhood of two to three percent. The remaining 97 to 98 percent of the electrical power applied to the input terminals of the loudspeaker is wasted as thermal power in heating up the voice coil windings and the magnet structure. What power is not wasted in heat is transformed into unwanted mechanical power.

So, by touting the "power handling capacity" of a loudspeaker, one is bragging about the relative inefficiency of the device. It's kind of like bragging that your automobile gets two miles to the gallon of gas.

A much more meaningful parameter for a loudspeaker is, how much acoustical power (sound pressure level, or SPL) does the device radiate for a given amount of electrical power applied?

To make a sound appear to be twice as loud, it requires ten times the amount of acoustical power or an increase of 10 dB SPL (1 Bel. [a decibel {dB} is 1/10th of a Bel]).

To put all of this into perspective, doubling the power output of a power amplifier will only cause a 3 dB increase in sound pressure level. Hence, to double the acoustical power (make the speaker twice as loud), you need a 10 dB change. To achieve this, you will need to have ten times the output power of your power amplifier.

In comparing the performance of any type of device, it is always prudent to question the efficiency of the devices under construction.

Properties of Sound

What we refer to as sound from a physical science sense is a disturbance created in air by some form of vibrating source. For example, if a firecracker explodes, air is "pushed out" in all directions from the point source. Another frequently used analogy is a pebble tossed into a pool of still water. When the pebble strikes the surface, ripples of waves are radiated as concentric circles from the point where the pebble caused the disturbance of the water.

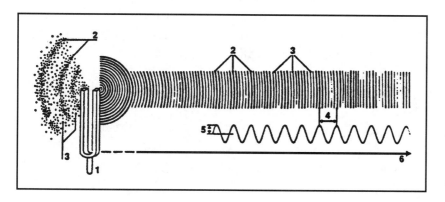

Figure 2-6. What sound might look like if a slice of air was made visible.

Air, like water, is an elastic and compressible medium, hence waves are formed by the disturbance. Wave crests are formed by compression of the molecules in the medium and wave troughs follow due to the relaxation of the compressed molecules into rarefactions.

Figure 2-6 with its schematic air molecules, represents what sound might look like if a slice of air was made visible.

The Greek letter (λ) lambda is used to denote wavelength. *Wavelength* is the distance between two sequential wave crests, troughs, or any other point we might choose to use.

One complete cycle would typically start at a point usually at zero and proceed up in pressure to maximum amplitude then back down to normal atmospheric pressure (zero disturbance), through zero to maximum negative amplitude, and thence back to zero. Continuous waves like these are described as having phases, which are like rotation, and thus are given numeration in degrees.

Velocity of Sound
Sound travels through some media faster than others. We usually think of sound as being in air, which is quite normal considering that humans exist in earth's atmosphere — which is air. However, whales, porpoises and other water dwellers use the medium of water to transmit sound. And, as can be seen from the following chart, sound has a velocity in water of five times what it has in air.

MATERIAL	SPEED OF SOUND	
GASES	(m/s)	(ft/s)
air	344	1,130
carbon dioxide	258	846
carbon monoxide	337	1,106
hydrogen	1,269	4,163
LIQUIDS		
alcohol	1,213	3,980
benzine	1,166	3,825

| turpentine | 1,376 | 4,515 |
| water | 1,435 | 4,708 |

SOLIDS

aluminum	5,104	16,745
brass	3,500	11,483
diamond	14,000	45,932
glass	5,500	18,045
steel	5,130	16,831
oak wood	3,850	12,631

Sound Velocity and Wavelength

The absolute sound velocity in air depends on the properties of the air, the temperature, and the pressure as the main considerations. For scientific measurements, the velocity value commonly used is 331.4 m/s (1087.4 feet per second) at a temperature of 0° C, with zero moisture content at a pressure of one atmosphere. For sound measurements the value used is ≈344 m/s, at a temperature of 20° C.

The speed of sound, in air, for any given temperature is given by the following equation:

$$V = V_{abs} \frac{(\sqrt{273} + t)}{16.52}$$

The wavelength of a given frequency may be calculated by dividing the velocity of sound in the medium being studied by the frequency in hertz:

$$2 = \frac{\text{velocity of sound}}{f}$$

So, for a sound wave of 1000 Hz propagated in air:

$$l = \frac{344}{1000} \text{ or } 0.344 \text{ m (1.13 ft)}$$

77

FREQUENCY	METERS	FEET
10 Hz	34.4	113
100 Hz	3.44	11.3
1000 Hz	.344	1.13
10,000 Hz	0.0344	0.113 (1.4 in.)

Intensity and Loudness

As noted above in the section under efficiency, sound has varying degrees of loudness. See *Figure 2-7*. The twice as loud theorem was developed by work that Alexander Graham Bell and his staff accomplished in the infancy of telephony. Prior to Bell's work there was a general appreciation within the scientific community that sound, distance, and loudness had some form of subjective relationship; but there was no quantification of the process.

The fact that the human ear is a logarithmic device really complicates the function. The good Doctor C.P. Boner used to say, "The good Lord made many wonderful and marvelous things; but He really messed up when it came to sound."

Not only does the human ear hear in a logarithmic fashion, it doesn't even respond to frequency in a linear fashion.

We perceive sound to be louder in the midrange and less intense at lower frequencies and higher frequencies (*Figure 2-8*). We also perceive "pitch and timbre" as a function of frequency and harmonics of a fundamental frequency. Also interesting is the fact that sound spans a range of approximately 12 octaves (scientifically, every time you double the frequency of a sound you go up one octave). In contrast, visible light spans one octave(!).

Octaves and Harmonics

When we relate frequency to music, we employ the use of the term *octave*.

The introduction of the term *octave* merely simplifies the concept by assigning audio frequencies to repeatable groups (scales) with

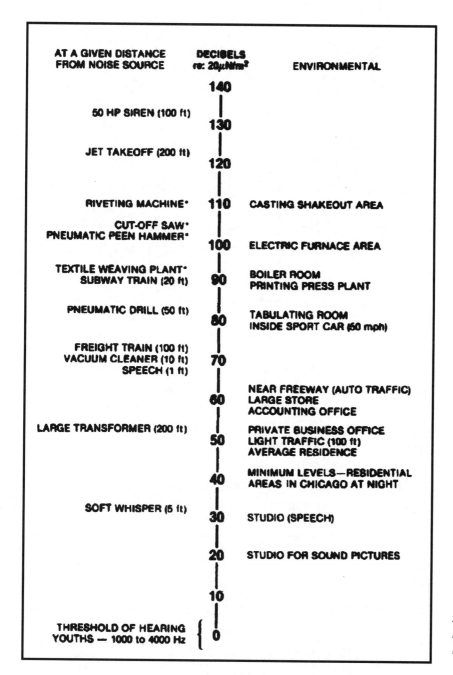

AT A GIVEN DISTANCE FROM NOISE SOURCE　　**DECIBELS re: 20μN/m²**　　**ENVIRONMENTAL**

140

50 HP SIREN (100 ft)

130

JET TAKEOFF (200 ft)

120

RIVETING MACHINE*　110　CASTING SHAKEOUT AREA

CUT-OFF SAW*
PNEUMATIC PEEN HAMMER*　100　ELECTRIC FURNACE AREA

TEXTILE WEAVING PLANT*
SUBWAY TRAIN (20 ft)　90　BOILER ROOM
PRINTING PRESS PLANT

PNEUMATIC DRILL (50 ft)　80　TABULATING ROOM
INSIDE SPORT CAR (50 mph)

FREIGHT TRAIN (100 ft)
VACUUM CLEANER (10 ft)　70
SPEECH (1 ft)

60　NEAR FREEWAY (AUTO TRAFFIC)
LARGE STORE
ACCOUNTING OFFICE

LARGE TRANSFORMER (200 ft)　50　PRIVATE BUSINESS OFFICE
LIGHT TRAFFIC (100 ft)
AVERAGE RESIDENCE

40　MINIMUM LEVELS—RESIDENTIAL
AREAS IN CHICAGO AT NIGHT

SOFT WHISPER (5 ft)　30　STUDIO (SPEECH)

20　STUDIO FOR SOUND PICTURES

10

THRESHOLD OF HEARING
YOUTHS — 1000 to 4000 Hz　0

Figure 2-7. Sound has varying degrees of loudness.

letter names (notes). Think of a piano keyboard as shown in *Figure 2-9.*

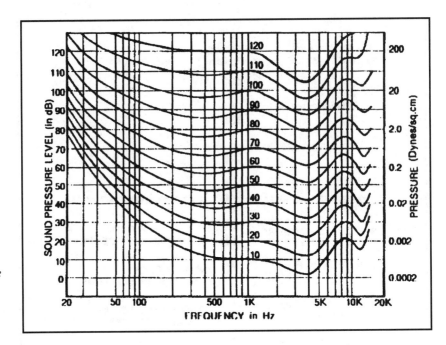

Figure 2-8. We perceive sound to be louder in the midrange and less intense at lower and higher frequencies.

The black-and-white key arrangement (12 per group) is such that this design repeats itself every octave through its 88 keys. Every C on a piano is at the same point in this repeating pattern of key arrangement. As you move up, or down, from one C to the next C, you are jumping one octave at a time.

The lowest fundamental frequency of C on a piano is 32.7 Hz, the next C (one octave higher) is 65.4 Hz, up to a high C of 4186 Hz.

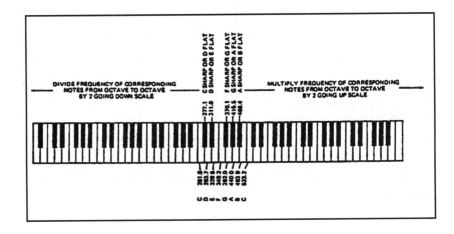

Figure 2-9. Piano keyboard as a model for octaves.

This concept of frequency doubling per octave applies to all of the notes in the musical scale.

The 88 keys on the piano span 7 octaves from 27.5 Hz to 4186 Hz.

Switching from the piano to a guitar for the moment, if someone plays the open low E of a guitar, the instrument produces a fundamental frequency of 82 Hz. However, it also produces a family of related frequencies which are known as harmonics of the fundamental. In the case of the low open E (82 Hz) additional frequencies will be produced at 164, 328, etc., hertz each at progressively lower sound pressure levels. Each harmonic is a doubling of the preceding frequency and are referred to as the second, third, fourth, etc., harmonic. Besides producing higher harmonics, the instrument can also produce sub-harmonics, i.e., frequencies which are half the fundamental, a quarter of the fundamental, etc.

It is the harmonics which gives the voice of either a singer or an instrument its color, character and tonal quality.

The same note played on different guitars will sound different because the harmonic content of that note will be different, because of the difference in design between guitar manufacturers. Two vocalists may be singing the same note, but because of the differences in their vocal cords, they produce different harmonic content.

A single note without any harmonics is known as a pure sine wave. Pure sine waves do not occur naturally in nature; only a flute, or a single note of the cardinal, comes close to sounding a pure tone. Only recently has the use of sine waves been employed in musical content. When used, they must be generated electronically by use of a synthesizer.

The Inverse Square Law
Sound and light are both radiation of the type commonly referred to as spherical radiation.

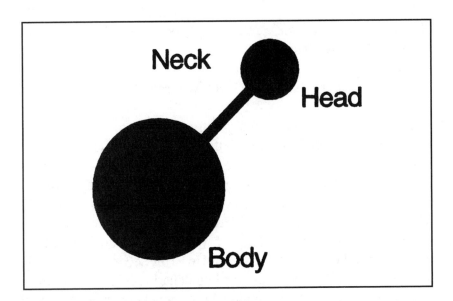

Figure 2-10. *A cow as two spheres.*

Physicists like to tell the joke about how to define a cow. If you asked four different people — say a farmer, a butcher, an accountant and a physicist — to describe a cow, you would likely get four very different answers. The farmer will likely see a cow as a unit of production, a butcher as parts of the whole, the accountant as dollars of production versus the futures market. However, the physicist is likely to start his definition by stating, "Think of a cow as a sphere." (*Figure 2-10*)

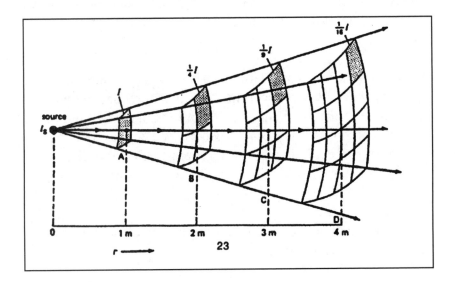

Figure 2-11. *Sound waves radiating from a nondirectional source emanate equally in all directions.*

Going back to the analogy of a pebble thrown into a pool, it will be observed that the concentric ripples are further apart and of lesser amplitude as they radiate outward from the source. Sound waves radiating from a nondirectional source emanate equally in all directions (see *Figure 2-11*).

From this graphic, we can see what is happening to the intensity of the sound wave as it progresses outward from the source.

The intensity will diminish as to the square of the distance from the source. Using the area "A" shown at reference distance "r" of 1 meter, we see that the formula for finding the area covered at various distances from the reference is:

$$Area = A \text{ x } r^2$$

Because the power of the sound is spread over a larger area, the intensity or level at each distance can be found by:

$$Intensity = 1/r2$$

— or the inverse of the square. Thus we have defined the inverse square law.

In the Level II handbook, we will be presented with the formula used to determine the change in a sound level when the distance from a sound source changes.

Properties of Light

Visible light which we, as humans can see, is produced by electromagnetic radiation or electromagnetic waves with a particular range of frequencies and wavelengths. Just as we perceive different audio wavelengths as different musical notes, we perceive light as a function of different wavelengths that represent the different colors in the visible spectrum of light.

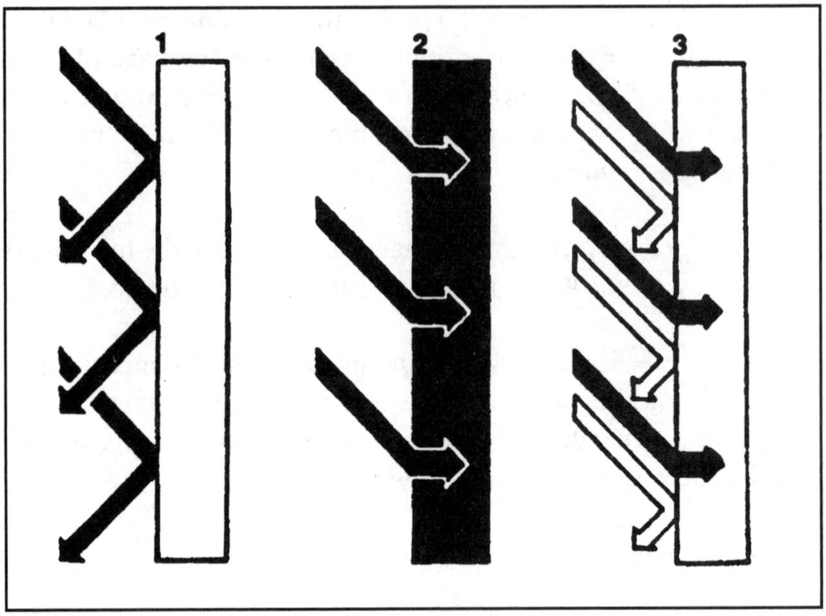

Figure 2-12. *The wavelengths reflected by a particular substance account for its color: (1) White reflects all colors; (2) Black absorbs all colors; (3) Red reflects red wavelengths and absorbs all others.*

The color spectrum spans 1 octave ranging from violet (3.9×10^{-7}m) to red (7.8×10^{-7}m). Ordinary or "white light" is a combination of all of the colors (wavelengths) that make up the visible spectrum.

Objects such as the sun, the stars (distant suns), or electric light bulbs produce their own light. Most of the objects around us do not produce light but are seen because they reflect light.

What gives objects their color? When white light strikes on a colored substance, most of these substances absorb some wavelengths in the spectrum and reflect others. The wavelengths a substance reflects account for its color.

1. A white substance reflects all the colors of the spectrum. (*Figure 2-12, #1*)

2. A black substance absorbs all the colors of the spectrum. (*Figure 2-12, #2*)

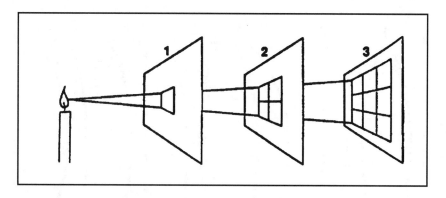

Figure 2-13. *A wedge of light focused through a series of windows.*

3. A red substance appears red because it reflects the spectrum's red wavelengths but absorbs all others. (*Figure 2-12, #3*)

As is true of all known spherical radiation, light intensity varies inversely with the square of the distance between the source and the illuminated surface.

Refer to *Figure 2-13* which represents a "wedge" of light focused through a series of "windows."

1. Some amount of light is falling onto a square.

2. At a distance that is twice as far from the source the same amount of light is spread over 4 squares.

3. At a distance that is x3 from the source, the light is spread over 9 squares.

Lenses
Lenses and other transparent substances change the direction of light by bending (refracting).

Some examples of various types of lenses are (*Figure 2-14*):

1. Double convex: used in magnifying glasses.
2. Plano-convex: used in some slide projectors.
3. Concave-convex: used to correct farsightedness.

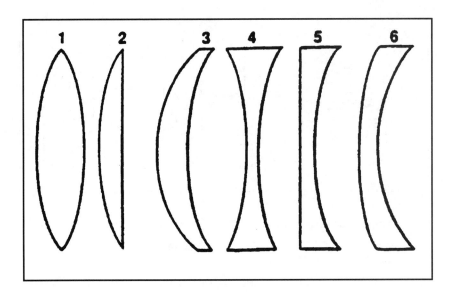

Figure 2-14. *Various types of lenses.*

4. Double concave: used to produce reduced images.
5. Plano-concave: used with other lenses in cameras.
6. Convex-concave: used to correct nearsightedness.

Concave Lens
A concave lens is thinner at the center than at the edges (*Figure 2-15*). It bends parallel light rays outward as they pass through the lens. A concave lens produces a reduced image that is the right way

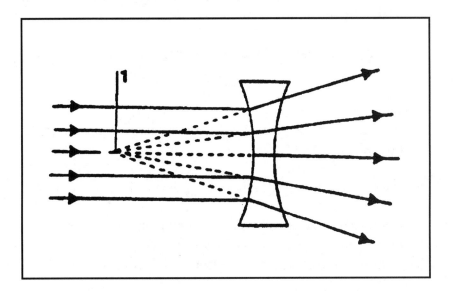

Figure 2-15. *Concave lens.*

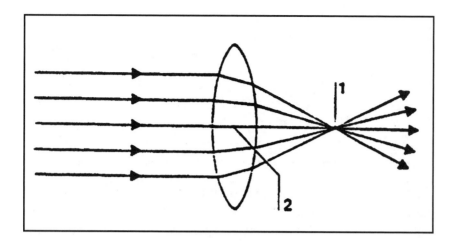

Figure 2-16.
Convex lens.

up and appears on the same side as the original object. This is a virtual (not real) image and it cannot be focused onto a screen.

Convex Lens

A convex lens is thicker at the center than at the edges (*Figure 2-16*). It bends light rays inward to meet at a focal point (1) behind the lens, producing a real image. The distance from the center of the lens (2) to the focal point is the focal length. An object that is less than one focal length from the lens becomes an enlarged image, the correct way up on the same side of the lens. If the object is moved further away a reduced inverted image is seen on the opposite side of the lens.

The Properties of Electricity and Magnetism

Electricity, magnetism and electromagnetic radiation are very much interrelated entities. A magnetic field cannot exist without an attendant electrical field and vice versa. Fortunately for us, this basic principle allows us to generate electricity, build electrical motors, employ transformers for stepping up and stepping down AC voltages, use relays and solenoids to control various switching applications, transmit radio and television signals, and a host of other electromagnetic functions.

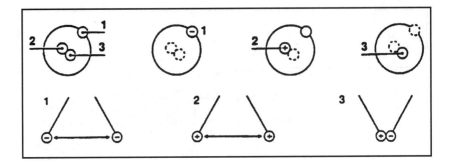

Figure 2-17. *(1)*
Electrons, (2) protons
and (3) neutrons, and
their relationships to
each other.

In both electricity and magnetism the behavior is a function of the atomic structure of the object upon which electrical or magnetic forces are imposed. Electricity flows through a conductor (wire) by virtue of the fact that some form of electromotive force (EMF) causes the electrons within the conductor to move from atom to atom in either a single direction (DC) or in an alternating direction (AC).

An *atom* is comprised of subatomic particles. For purposes of this discussion we will confine our discussion to those particles which affect the conduction of electricity and the creation of magnetic fields. Subatomic particles which concern the nuclear and/or theoretical physicist can probably be ignored without damage to this discussion (*Figure 2-17*).

1. *Electrons* are negatively charged (-).

2. *Protons* are positively charged (+).

3. *Neutrons* are neutral; they have no charge.

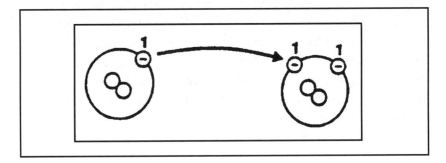

Figure 2-18.
Electricity.

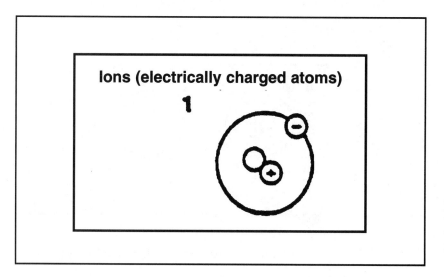

Figure 2-19. *The number of electrons and protons is equal in neutral atoms.*

a. Negatively-charged particles repel each other.
b. Positively-charged particles repel each other.
c. Oppositely-charged particles will attract each other.

Electricity is energy which is released when negatively charged electrons move from atom to atom (*Figure 2-18*).

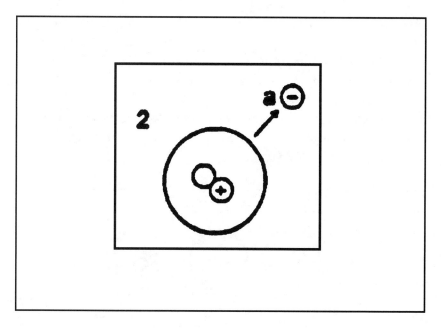

Figure 2-20. *Removing an electron from a neutral atom produces a positively charged ion.*

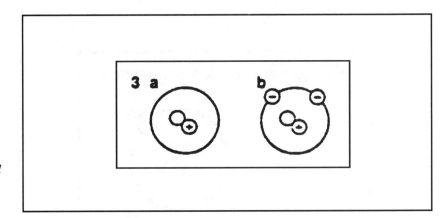

Figure 2-21. *The number of protons and electrons in an ion is different.*

In electrically neutral atoms the number of electrons and protons is equal (*Figure 2-19*).

Removing an electron from an electrically neutral atom produces a positively charged *ion* (*Figure 2-20*).

In an ion, the number of electrons and protons is different (*Figure 2-21*). When there are more protons, it is a positive ion. A negatively charged ion has more electrons than protons.

An electrical *current* is a continuous flow of electrons through a conductor such as a copper wire when an electromotive force is applied to the conductor. If no current is flowing, the charge is at rest (*Figure 2-22*).

In *direct current* (DC) circuits, electrons move in the conductor in one direction only, between the negative and positive terminals of the circuit (*Figure 2-23*).

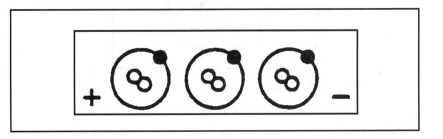

Figure 2-22. *Electrical current is a continuous flow of electrons through a conductor.*

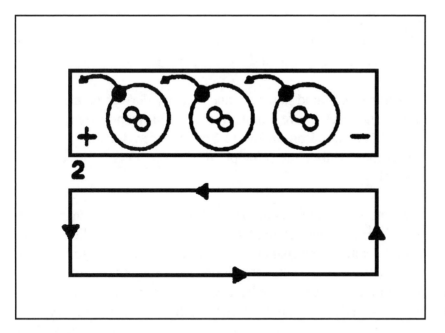

Figure 2-23. *Direct current (DC).*

In *alternating current* (AC) circuits, electrons move in the conductor in an alternating way, first one way then the other (*Figure 2-24*).

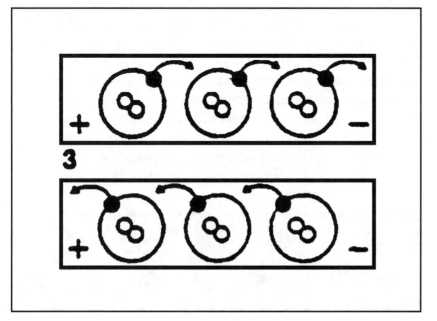

Figure 2-24. *Alternating current (AC).*

Properties of Magnets

Magnetism is a force of attraction or repulsion between certain materials, particularly iron and some other metals.

Magnetism is found naturally and actually derives its name from Magnesia which is a region in the middle East where lodestones (a natural magnetic iron ore) were first discovered in antiquity. Any object that exhibits the principle of magnetism is referred to as a magnet.

Magnets as we know them can be classified as permanent magnets and electromagnets. A hard steel or steel alloy bar when magnetized becomes a permanent magnet because it tends to retain its magnetism under normal conditions for a long period of time unless subjected to excessive heat or jarring. The condition of retaining a magnetic charge is referred to as the *permeability* of the magnet. Cores for electromagnets are generally made of soft iron which accepts a magnetic charge quickly; but due to its low permeability also loses its charge very quickly.

A piece of material that is susceptible to being permanently magnetized is called a *ferromagnetic* metal (see *Figure 2-25*). A mass of ferromagnetic material consists of molecules which act like miniature magnets that are called *dipoles*. In an unmagnetized condition these dipoles form closed links and following the atomic theory of attraction and repulsion of like and unlike particles, the north-seeking and south-seeking (north and south) poles of each of the particles attract and there are no FREE poles.

When the bar of ferromagnetic material becomes magnetized, the closed chains of dipoles are broken and aligned as illustrated in

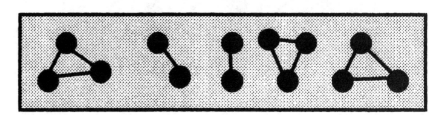

Figure 2-25.
Ferromagnetic metal.

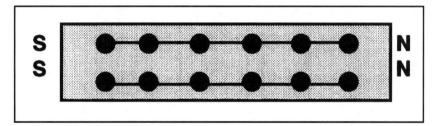

Figure 2-26. *When ferromagnetic material is magnetized, the closed chains of dipoles are broken and aligned as shown.*

Figure 2-26. By virtue of this action two FREE poles are created at either end of the bar.

An iron bar magnet as shown in *Figure 2-27* sets up a magnetic field which is represented by the curved lines flowing between the poles. By convention, these lines are said to flow from the north pole to the south pole of the same magnet or another magnet within the magnetic field. The density of these imaginary lines converges at the poles of the bar and the magnetic radiation is inversely proportional to the distance from the poles.

During the 18th century C.A. Coulomb (French physicist 1736 - 1806) discovered what is known as Coulomb's law which states

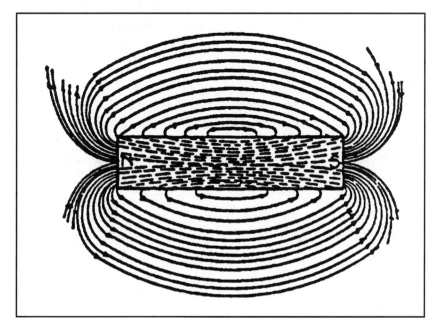

Figure 2-27. *An iron bar magnet.*

that the force of attraction or repulsion between two magnetic poles is directly proportional to the product of the strengths of the poles and inversely proportional to the square of the distance between them. And, once again we encounter the same inverse square law that governs the properties of sound.

The lines which represent the magnetic field of a magnet are commonly known as "lines of magnetic induction." All of the lines together represent the magnetic *flux*, and this gives rise to the expression magnetic flux density. As the density increases the strength of the respective poles increase.

One line of the flux is called a maxwell (M). The flux per unit area, or the number of maxwells per square centimeter is the *flux density* which is designated (B). The unit of flux density, or one maxwell per square centimeter was previously referred to as a *gauss*. Under the SI unit the designator for flux density is the *tesla* (T). 1T = 10,000 gauss.

One of the places where you will run across the use of flux density expressed in teslas is in the specifications for a loudspeaker. In fact, you will also find cases where the flux density is stated in the older term gauss. The flux density of a loudspeaker is an important parameter in the design and performance of a loudspeaker's engine.

Lines of magnetic induction are always closed loops. Each line, or loop, may be thought of as acting within itself somewhat like a stretched rubber band in that it tends to become as short as possible. Yet each of these lines has a repelling effect upon its neighbors, tending to keep them separated from each other.

To test this theory of magnetic flux, try placing a plate of glass on a bar magnet and then sprinkle the glass with ferromagnetic iron filings. The iron filings will arrange themselves as illustrated in *Figure 2-28.*

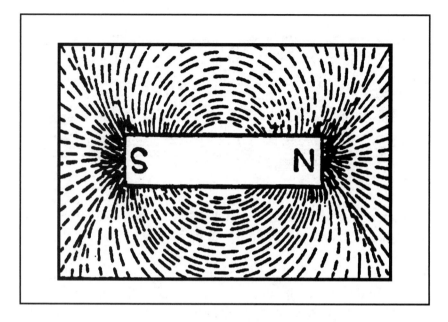

Figure 2-28.
Magnetic flux.

If a second magnet is placed at the end of the first magnet so that their north and south poles are attracted, the magnets will become additive (*Figure 2-29*).

If the two magnets are placed in position so that the poles are repelling, the two fields are opposing (*Figure 2-30*).

Since there is a stronger magnetic field nearest the poles, we can create a field of greater intensity by bending the bar into the form

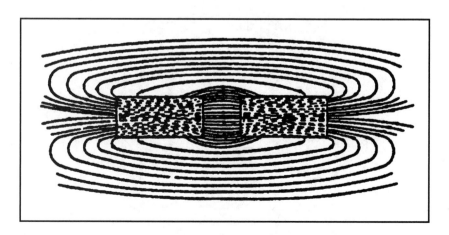

Figure 2-29.
Additive magnets.

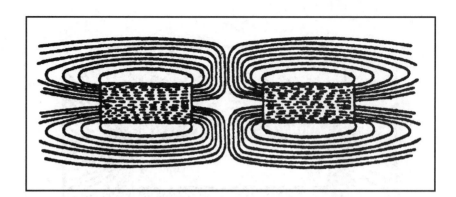

Figure 2-30.
Opposing magnetic fields.

of a horseshoe (*Figure 2-31*). The field of a horseshoe magnet is more intense than that of a single bar having the same pole strength.

This gives us, for a magnetic circuit, an analogy to an electrical circuit. In the electrical circuit, if we connect a conductor between the positive and negative terminals of a battery and decrease the length of the conductor we decrease the resistance and increase the current strength.

In the case of the magnet circuit, when we decrease the length of the path from the north to the south pole; we increase the number of lines of magnetic induction.

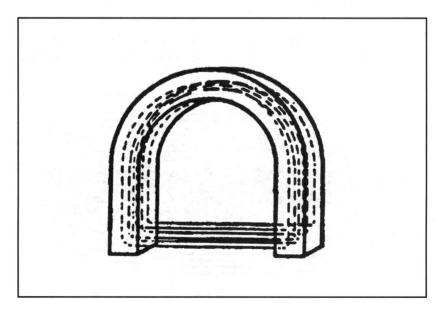

Figure 2-31.
Magnetic field of a horseshoe magnet.

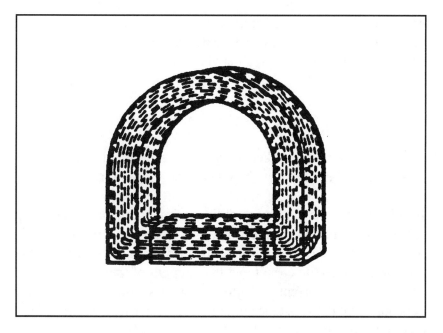

Figure 2-32. Increase in magnetic induction due to magnetic material added between the horseshoe poles.

Now, if we insert a piece of soft iron or other magnetic material between the poles of the horseshoe, we again greatly increase the number of lines of magnetic induction existing in the circuit formed by the magnet itself and the soft iron used for closing the circuit between the poles (*Figure 2-32*).

The magnetic circuit described above is analogous to decreasing the resistance of an electrical circuit by substituting a conductor of lower resistance in place of one having a higher resistance.

If we construct our magnet so that the south pole is the inner core of a concentric circle and the north pole is the outer ring of the concentric circle, we have, in essence, the magnet structure of a loudspeaker engine. The gap between the two poles would be where the voice coil is positioned. In reality, the gap for the loudspeaker assembly is kept very small; hence, the flux density is exceedingly high.

Just as electrical current can be caused to flow in an electrical circuit, magnetic flux can be established in a magnetic circuit.

Magnetic flux *f*, or the total number of lines of induction existing in the circuit, is in some respects analogous to electric current. The flux density (β) per unit area may be written:

$$b = \frac{f}{A}$$

— where *A* = the area taken at right angles to the direction of the flux, and *f* = the flux through and normal to this area.

It follows from the discussion of lines of magnetic induction being increased by the insertion of materials other than air in the magnetic field, that the flux density depends upon the materials of the completed magnetic circuit and the strength of the magnet. In the same sense, the current strength in a cross section of a conductor depends upon the resistance of the closed electrical circuit and the electromotive force applied.

We may then consider that there is a property of the magnetic circuit which is analogous to the resistance of an electrical circuit. This property is called *reluctance* (R).

Likewise, there is a property of the magnet which is analogous to the electromotive force of a battery. The magnetic force is called the *magnetomotive force* (M) and is expressed in "gilberts."

For a completed magnetic circuit, we can apply an equation which is identical in form to Ohm's law:

THE FLUX FOR ANY GIVEN MAGNETIC CIRCUIT IS EQUAL TO THE MAGNETOMOTIVE FORCE OF THE MAGNET DIVIDED BY THE RELUCTANCE OF THE CLOSED CIRCUIT.

Compare: $f = \dfrac{M}{R}$ and $I = \dfrac{E}{R}$

— where the symbol for flux is *f*, for magnetomotive force *M* and reluctance *R*.

In practice the above magnetic equation is seldom used in the form shown but from this relation we derive other equations dealing with flux density, field intensity and the magnetic properties of iron.

While we can draw analogies between a magnetic circuit and an electric circuit, the analogy is not complete since there are ways in which the two circuits differ. The two more important differences are:

1. A magnetic circuit can never be entirely opened; a magnetic field must exist at all times in the vicinity of a magnet.

2. Flux is not strictly analogous to current since current is rate of flow of electrical charge while the nature of flux is more nearly a state or condition of the medium in which it is established.

Electromagnets
When an electric current is passed along a wire, a magnetic field is created around it.

To illustrate this principle, refer to *Figure 2-33*, where a vertical conductor carrying an electrical current pierces a piece of cardboard on a specific plane. A magnetic field with lines of magnetic induction can be detected encircling the conductor.

To illustrate further, if ferromagnetic filings are sprinkled on the cardboard they will align, as shown in *Figure 2-34*, and will form visible concentric circles.

These observations allow us to state: whenever an electrical current is flowing, there is an established magnetic field; and the loops formed by the encircling lines are always in a plane perpendicular to the electrical conductor.

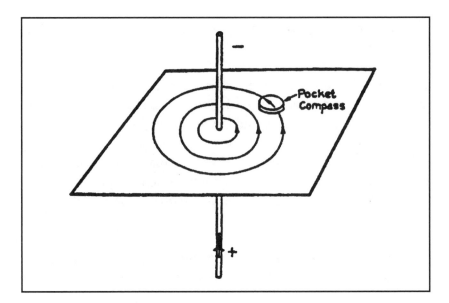

Figure 2-33. A vertical conductor carrying an electrical current pierces a piece of cardboard on a specific plane.

The magnetic field formed around a single conductor is by itself a relatively weak field. If, however, we form the conductor into a loop, the magnetic intensity inside of the loop is greatly magnified (*Figure 2-35*). Furthermore, if we construct a circuit consisting of several loops, or turns, of wire on a coil, the strength of the field is multiplied by the number of turns.

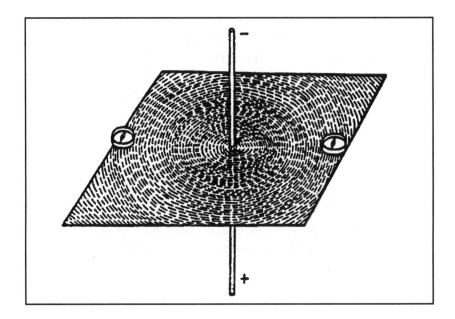

Figure 2-34. If ferromagnetic filings are sprinkled on the cardboard, they will align as shown.

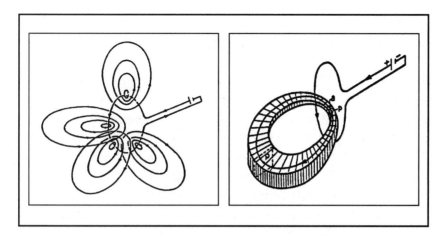

Figure 2-35. Forming the conductor into a loop increases the magnetic intensity inside the loop.

If we then further add a soft iron core to the coil, the number of lines of magnetic flux will again be increased.

The next step is to increase the electrical current strength through the coil, and we will find that the intensity of the magnetic field is directly proportional to the increase in the electrical current.

The coil of wire shown in *Figure 2-36* is referred to as a *solenoid*. Inasmuch as the magnetic intensity is proportional to the number

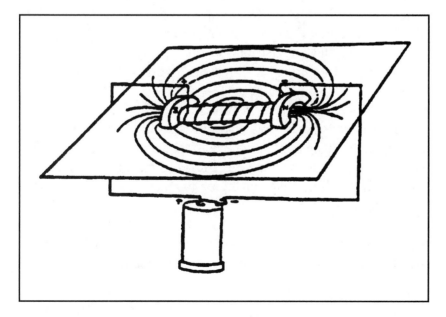

Figure 2-36.
A solenoid.

of turns, then the longer the coil is for a given diameter and a given current flow, the stronger the magnetic density.

Any discussion of the properties of magnetic coils usually begins with an analysis of the coil, assuming an air core, and proceeds to what conditions vary when a ferromagnetic core is substituted for the air core. As previously noted, introducing a soft iron core reduces the reluctance of the circuit. It is more common to use the reciprocal of reluctance to denote magnetic conductivity. The unit of magnetic conductivity is called *permeability* and is represented by the Greek letter μ. Permeability is the ratio of the magnetic conductivity of a substance to the magnetic conductivity of air.

Permeability has been compared to electric conductivity. However there is an essential distinction — the stability of iron under various degrees of magnetization is not equal to that of the ordinary metallic electrical conductor. In an electrical circuit the resistance or conductivity remains relatively fixed for any degree of current strength. With an air core coil the analogy is by and large similar; however with a ferromagnetic core when the flux density is increased the permeability of the iron is changed. And, at some point the core will become saturated and a further increase in the magnetizing force will not produce a proportional increase in flux density. This condition is of extreme importance when designing and installing loudspeaker distribution lines using step-up and line-matching transformers.

Properties of Electromagnetic Waves
Electromagnetic radiation is energy resulting from the interrelationship between an alternating electrical circuit and the attendant magnetic field. If a changing electrical current can create a magnetic field, it follows that a changing magnetic field can create an attendant electrical field.

Is an audio signal an electromagnetic field? As we learned earlier, sound is the movement of molecules within a medium (usually air).

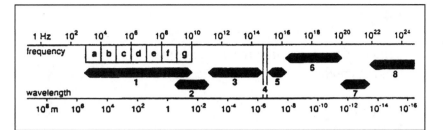

Figure 2-37. Various forms of radiant energy, arranged in order of frequency and wavelength to produce the electromagnetic spectrum.

Electromagnetic waves can transverse a vacuum, hence an audio signal is not an electromagnetic radiation.

The theory of electromagnetic radiation was developed by James Maxwell (Scottish physicist 1831-1879) and published in 1865. Maxwell hypothesized that the speed of propagation of electromagnetic radiation should equal the speed of light. Further experimentation by Heinrich Hertz (German physicist 1857-1894) verified Maxwell's predictions and Hertz went on to identify the existence of radio waves (also referred to as Hertzian waves).

The various types of electromagnetic radiation differ only in wavelength and frequency, otherwise they are alike in all respects. The propagation rate of all such waves is 186,282 miles per second (299,792 km per sec). The wavelength of electromagnetic waves, like audio waves, is a function of frequency and is expressed by the Greek letter lambda (l).

Figure 2-37 shows the various forms of radiant energy, arranged in order of their frequency and wavelength to produce the electromagnetic spectrum.

1. Radio waves: radio and television broadcasting.
2. Radar and microwaves.
3. Infrared waves: radiant heat.
4. Visible light.
5. Ultraviolet light.
6. X-rays.
7. Gamma rays.
8. Cosmic rays.

Questions

1. Which of the following mechanical quantities are not fundamental?

   a. Volume
   b. Time
   c. Mass
   d. Area
   e. Length

2. What force (on earth) causes an object to have weight?

3. If 1 cu. ft. of water weighs » 62.5 lbs., what is the weight of 1 cu. m of water?

4. If a boulder is at rest on the surface of the earth, does it possess any potential energy in kilograms?

5. A power amplifier is rated at 200 watts (at full power output). It requires 110 volts at 2.4 amperes of line current. What is the efficiency of the amplifier?

6. What is the commonly used value for the speed of sound in air at 20° C?

7. What is the wavelength of an audio frequency tone of 5 kHz?

8. What is the second harmonic of A above middle C?

9. Define the term "inverse square law."

10. What is the result if the current applied to an air-core electromagnet is doubled?

(Answers in Appendix A.)

DC CIRCUITS

Sources of DC Current

The most common source of DC energy (EMF, or electromotive force) is the battery. A battery converts stored chemical energy into electrical energy.

Consider your automobile's battery. Schematically, the wet battery is shown in *Figure 2-38*:

1. Each cell of the battery has a liquid electrolyte, which serves as the medium of transfer for the electrons that migrate between electrodes.

2. The electrochemical process of the battery causes the positive and negative-charged ions to migrate to the electrode of opposite charge. A reaction occurs at the negative electrode, allowing electrons to flow into the electrode from an outside source and be consumed. While a reaction at the positive electrode produces electrons that can flow into the external circuit.

3. If an external conductor is connected between the positive and negative terminals of the battery, then current will flow through the external conductor.

A dry battery, such as you would use in a flashlight or transistor radio, works in a similar matter to a wet cell battery. The dry cell has a paste electrolyte instead of a liquid electrolyte.

The parts of a dry cell are illustrated as shown in *Figure 2*-38. The parts of the battery are:

1. Paste electrolyte.
2. Positive anode.

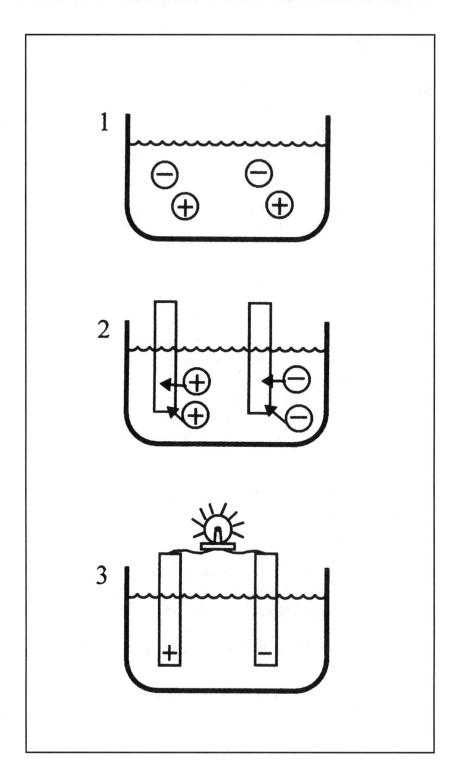

Figure 2-38. A wet battery: (1) cell with liquid electrolyte, (2) electrochemical process, and (3) external conductor.

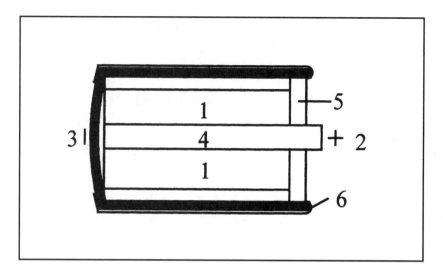

Figure 2-39. Parts of a dry cell battery (see text).

3. Negative anode.
4. Carbon rod.
5. Insulating seal.
6. Insulating casing.

The Property of Resistance

All matter has, to some varying degree, a resistance to the flow of electrical current. Or said another way, various materials, when exposed to an electromotive force of the same magnitude, exhibit a greater movement of electrons through their atomic structure than some others.

Materials such as copper, water, iron and silver have a low resistance to the flow of electrical current and are referred to as conductors. Conversely, materials such as wood, glass and rubber have a strong resistance to the flow of electrons and are referred to as insulators.

The following charts indicate the relative degree of conductivity and insulating properties:

Resistivity of Various Metal Conductors at 0° C
(Compared to pure copper of the same length and cross-sectional area.)

TYPE OF METAL	TIMES THE RESISTANCE OF PURE COPPER
Silver	.941
Copper (pure)	1.000*
Copper (hard drawn)	1.025
Gold	1.423
Aluminum	1.679
Magnesium	2.788
Zinc	3.449
Tungsten (hard drawn)	3.474
Nickel	4.442
Iron (pure)	5.673
Platinum	6.301
Tin	6.730
Steel (soft)	7.564
Tantalum	9.359
Lead	12.692
German Silver	21.218
Steel(hard)	29.294
Mercury	60.301
Cast Iron (hard)	62.692

*Resistivity of pure copper = 1.7×10^{-8} ohms-meter.

INSULATING MATERIALS
(Given in the order of their approximate insulating properties, from the most to least.)

Dry air	Rubber	Dry wood
Shellac	Porcelain	Slate
Paraffin	Sulphur	Fiber
Paraffin paper	Glass	Distilled water

Paraffin oil   Varnish   Alcohol
Ebonite   Celluloid

Direct Current (DC) Calculations

The solution of DC circuits relies on the principles expressed in Ohm's law. The understanding of this law will allow you to solve almost any problem encountered in any DC circuit analysis.

Ohm's Law
Ohm's law is probably the most fundamental law governing the interaction of electrical energy. It also illustrates the prime necessity for understanding algebra and the manipulation of terms in algebraic notation.

Ohm's law is named after the German physicist Georg Simon Ohm (1787-1854), who discovered the relationship between voltage, current, and resistance in an electric circuit. It is expressed as:

For any circuit or part of a circuit under consideration, the current in amperes is equal to the potential difference in volts divided by the resistance in ohms.

This law, mathematically expressed, is as follows:

$$Current = \frac{Potential\ Difference}{Resistance} \text{ or } Amperes = \frac{Volts}{Ohms}$$

If in the above expression we substitute the proper symbols for the units of current, potential difference and resistance (or instead of amperes, volts and ohms), we have the following equation:

$$A = V \div \Omega \text{ or } A = V/\Omega$$

If $A = V/\Omega$ then $V = A \times \Omega$ and $\Omega = V/A$

The relationship between resistance, potential difference and current results in the conversion of energy from chemical to heat, or some other form. These expressions therefore have a definite power consumption or represent a definite transfer of power to some external device. The scientific unit for work is the joule (J), and the scientific unit for the rate of work or power is the watt (W).

In the electrical circuit, if we multiply the potential difference in volts by the current in amperes, we have an expression for the *power* in watts.

The watt, therefore, can be defined as an electrical unit as well as a mechanical unit. It is the power expended in a circuit having a potential difference of one volt and a current of one ampere.

Because the watt is the connecting relationship between mechanical units and electrical units, the following formula should be considered second only to Ohm's law in importance:

Power = Potential Difference x Current, or W = VA

Ohm's law states that $V = A\Omega$. Therefore, if we substitute $A\Omega$ for V in the above equations, we can write:

$W = (A\Omega)(A)$ or $W = A^2\Omega$

Similarly, if W = VA, then substituting for $A(A = V/\Omega)$ in the formula gives us:

$W = (V)(V/\Omega)$ or $W = V^2/\Omega$

The other expressions of Ohm's law are similarly algebraic derivatives of the base formula.

The pie chart in *Figure 2-40* shows the relationship between the electrical quantities that are components of Ohm's law.

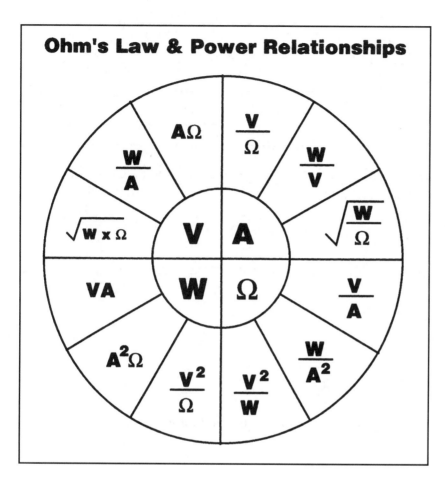

Figure 2-40. The relationship between the electrical quantities that are components of Ohm's law.

Resistors in Series

Two or more resistors connected as shown in *Figure 2-41* are said to be connected in series. The equivalent resistance of a set of series-connected resistors is that value of resistor that can be substituted for the entire set without changing the properties of the circuit.

The potential difference V across the set is equal to the sum of the potential differences V_1, V_2 and V_3 across each individual resistor. Hence:

$$V = V_1 + V_2 + V_3$$

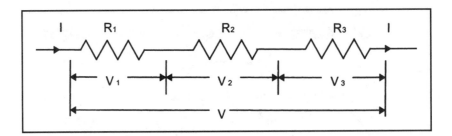

Figure 2-41.
Resistors connected in series.

Because the current in each resistor is I, the potential differences across them are:

$$V_1 = IR_1 \, , \, V_2 = IR_2 \, , \, V_3 = IR_3$$

And the potential difference across the equivalent resistance R is:

$$V = IR$$

Substituting for V gives:

$$IR = IR_1 + IR_2 + IR_3$$

Dividing both sides of the equation by I gives:

$$R = R_1 + R_2 + R_3$$

Consequently, we can write that the equivalent resistance of any set of series-connected resistors is equal to the sum of the individual resistances.

$$R = R_1 + R_2 + R_3 + R_n$$

Questions

1. When limiting the current in a 50Ω resistor to 10A when it is connected to a 600V power supply, what should be the value of the current limiting series resistor?

2. What is the voltage drop across each of the individual resistors in the above question?

3. Two loads are connected in series across a 12V source. One of the loads has a resistance of 5Ω and the other is 10Ω.

    a. What is the current in each load?
    b. What is the voltage drop across each load?
    c. What powers are dissipated by each load and by the combination?

4. A 2 kΩ and a 5 kΩ resistor are in series. A voltmeter shows the potential difference across the 2 kΩ resistor to be 2V. What is the current in each resistor and the potential difference across the 5 kΩ resistor?

(Answers in Appendix A.)

Voltage Dividers

In most cases, an electronic device (a power amplifier for example) is powered from an AC line source. The AC is then rectified and filtered to provide DC to the device. The output of the DC power supply is usually a voltage that corresponds to the largest voltage required by the device, and a voltage divider is used to obtain the other voltages that might be required to operate the device.

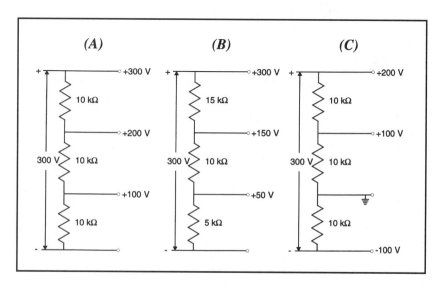

Figure 2-42. *Typical voltage dividers encountered in a DC power supply.*

Figure 2-42 represents three different, typical voltage dividers that you might encounter in a DC power supply.

A voltage divider may be a set of resistors connected in series, or it might be one resistor "tapped" at various points.

In *Figure 2-42(a)*, all three resistors are of equal value and the open circuit voltage across each of the three legs are one-third of the power supply voltage.

In *Figure 2-42(b)*, the resistor values are different, hence the open-circuit voltage values will differ; however, they will add up to the value of the power supply voltage.

In *Figure 2-42(c)*, the circuit provides both positive and negative voltages by virtue of being grounded at an intermediate point rather than at one end of the series of resistors.

When these voltage dividers are connected to loads, the voltages at their intermediate terminals will differ from the open circuit voltage depending upon the current flow drawn by the load.

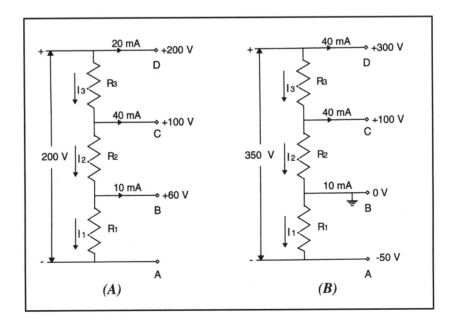

Figure 2-43. *Voltage divider examples.*

Questions

1. A voltage divider (*Figure 2-43[A]*) is to provide currents of 20 mA (1 mA=0.001A, so 20 mA = 0.02A) at 200V, 40 mA at 100V, and 10 mA at 60V as diagrammed below.

    Find the value of the three resistors required.

    Assume that the current through R_1 (the bleeder current) is 10% of the total of the load currents, which is a common proportion.

2. Find the values of the resistors in the voltage divider for *Figure 2-43(B)*. The -50V terminal is a bias voltage and no current is drawn. The bleeder current I_2 is 8 mA.

(Answers in Appendix A.)

Resistors in Parallel

Figure 2-44 is representative of a simple parallel circuit.

Using Ohm's law to solve for either of these parallel circuits, we find that the current in either branch must be equal to the potential difference measured across the particular resistance divided by its value in ohms; and for the particular circuit, the potential differ-

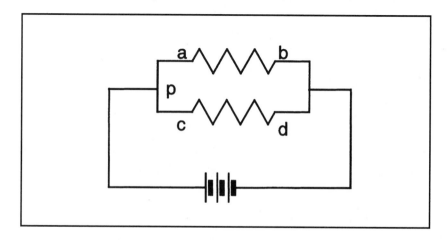

Figure 2-44. *A simple parallel circuit.*

ence measured across either resistance is the potential difference of the battery. The battery, in reality, is supplying two currents, one through the resistance ab and the other through the resistance cd.

The two parallel branch currents are united and flow together in the conductors connecting the poles of the battery with the junctions of the two parallel resistances. Consequently, for any circuit having two or more parallel resistances, the current supplied to the combination must be greater than the current supplied to either of the individual resistances.

If we think of the combination of resistances shown in *Figure 2-44* as being equivalent to a single resistance that might be substituted in their place, we can state that the value in ohms of two resistances in parallel is less than that of either resistance taken singly.

The calculation for the solution of parallel resistances is aided by an understanding of Kirchoff's First Law which states:

At any point in a circuit there is as much current flowing to the point as there is away from it.

Kirchoff's First Law applies regardless of the number of branches that may be connected to the point in question.

The law can be interpreted by its application to point P in the above parallel resistance network. If I is the current being supplied by the source to the combination of the two resistances in parallel, and I_1 and I_2 are the respective currents through the two parallel resistances, then:

$$I = I_1 + I_2$$

Applying Ohm's law to the entire circuit and with R represent the value of the combined resistances in parallel, we have:

$$R = E /I \text{ or } R = E/(I_1 + I_2)$$

However:

$$I_1 = E/R_1 \text{ and } I_2 = E/R_2$$

Therefore:

$$R = E/((E/R_1) + (E/R_2))$$

By canceling the E's in the above equation we have:

$$R = 1/((1/R_1) + (1/R_2))$$

Simplifying we have:

$$R = R_1 \times R_2/(R_1 + R_2)$$

Thus we have a general equation for calculating the combined value of any two parallel resistances. This may be stated as:

To obtain the combined resistance of any two resistances in parallel, divide their product by their sum.

A similar proof can be established for the calculation of any number of parallel resistors and we can write:

$$1/R = 1/R_1 + 1/R_2 + 1/R_3 + 1/R_n...$$

We can also write the above formula as:

$$R = 1/(1/R_1 + 1/R_2 + 1/R_3 + 1/R_n...)$$

The use of a calculator with a reciprocal (1/X) key makes the calculation of the above expression relatively simple to accomplish. However, the calculator being used in class uses the notation of X-1 for finding a reciprocal. So, we must rewrite the above formula to the new notation. It becomes:

$$R = (R_1^{-1} + R_2^{-1} + R_3^{-1} + R_n...^{-1})^{-1}$$

This notation allows you to enter the whole formula in one set of keystrokes without having to find intermediate answers.

The X^{-1} key is a second function of the X^2 key. You must use the *2nd F* key to access the X^{-1} function.

Questions

1. Find the equivalent resistance of a 5Ω and a 10Ω resistor connected in parallel.

2. A circuit has a resistance of 50Ω. How can it be reduced to 20Ω?

3. Two 240-ohm light bulbs are to be connected to a 120V power source. To determine whether they will be brighter when connected in series or in parallel, calculate the power they dissipate in each arrangement.

4. Refer to *Figure 2-45*. A potential difference of 60V is applied to this circuit. What is the current in each resistor and the total current in the circuit?

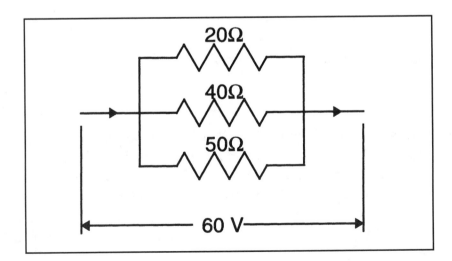

Figure 2-45. *Circuit for Resistors-in-Parallel Question #4.*

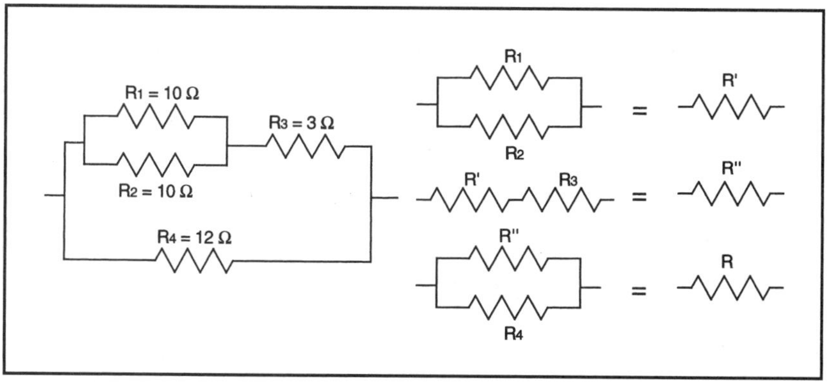

Figure 2-46. *A resistive network with both series and parallel resistors.*

(Answers in Appendix A.)

Resistors in Series/Parallel Networks

Figure 2-46 illustrates a resistive network having both series and parallel resistors. The procedure to solve such a network involves calculating the parallel components first, substituting the equivalent value of a series resistor, and then solving for the total resistance of the circuit. In more complex series/parallel networks this procedure might have to be repeated several times.

Questions

1. Solve *Figure 2-46* for the equivalent resistance of the circuit.

2. A potential difference of 20V is applied to *Figure 2-47*. Find the current through each resistor and the current throughout the entire circuit.

(Answers in Appendix A.)

Resistance and Wire Size

In addition to the law showing the relation between electromotive force, current, and resistance, Ohm investigated the properties of

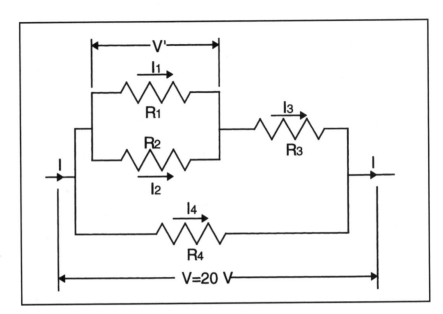

Figure 2-47. *Circuit*
for Resistors in Series/
Parallel Networks
Question #2.

conductors and established in addition to their relative values the following laws:

1. The resistance of any conductor varies directly with its length.

2. The resistance of any conductor varies inversely with its cross-sectional area.

A conductor connected between the positive and negative terminals of a DC source (a battery) has some degree of resistance which is determined by:

1. The length of the wire between the positive and negative terminals.

2. The cross-sectional area of the conductor used.

3. The material out of which it is constructed.

4. The ambient temperature to which the conductor is exposed.

A circuit of this type can be considered as the most elementary form of series resistance. The following relationships will prevail.

1. The resistance of a conductor is directly proportional to its length:

$$R_1/R_2 = L_1/L_2$$

2. The resistance of a wire varies inversely with its cross-sectional area A. All things being equal the thicker a wire is, the less its resistance:

$$R_1/R_2 = A_2/A_1$$

The discussion of specific resistance, conductive property of a material, and change of material conductivity due to ambient temperature is beyond the scope of Level I and will be dealt with in subsequent chapters.

It is common practice in electrical and electronic work to express the area of a round conductor in terms of circular mils. This avoids the necessity to multiply and divide by π. The *mil* is a unit of length equal to 0.001, (1×10^{-3}), $(1/1000)$ inch. A circular mil is a unit of area equal to the area of a circle whose diameter is 1 mil. The area A_{cmil} in circular mils of a circle whose diameter in mils is D_{mil} is equal to D^2mil:

$$A_{cmil} = (D_{mil})^2$$

United States manufacturers of wire size their product in accordance with the American Wire Gauge (AWG) system (*Table 2-1*). In this system of measurement, the largest wire is AWG No. 0000 (460 mils in diameter - almost 1/2") and the smallest is No. 40 (3.1 mils - hair thin sized).

The sequence of AWG wire gauge is such that every third gauge number means a cross-sectional area half as great and hence a re-

American Wire Gauge Table
Resistance of bare annealed copper wire at 20°C (68°F)

AWG No.	Diameter D, Mils	Area D², cmil	Resistance Ω/1000 ft
0000	460.0	211,600	0.0490
000	409.6	167,800	0.0618
00	364.8	133,100	0.0779
0	324.9	105,500	0.0983
1	289.3	83,690	0.1239
2	257.6	66,360	0.1563
3	229.4	52,630	0.1970
4	204.3	41,740	0.2485
5	181.9	33,100	0.3133
6	162.0	26,250	0.3951
7	144.3	20,820	0.4982
8	128.5	16,510	0.6282
9	114.4	13,090	0.7921
10	101.9	10,380	0.9989
11	90.74	8,234	1.260
12	80.81	6,530	1.588
13	71.96	5,178	2.003
14	64.08	4,107	2.525
15	57.07	3,257	3.184
16	50.82	2,583	4.016
17	45.26	2,048	5.064
18	40.30	1,624	6.385
19	35.89	1,288	8.051
20	31.96	1,022	10.15
21	28.46	810.1	12.80
22	23.35	642.4	16.14
23	22.57	509.5	20.36
24	20.10	404.0	25.67
25	17.90	320.4	32.37
26	15.94	254.1	40.81
27	14.20	201.5	51.47
28	12.64	159.8	64.90
29	11.26	126.7	81.83
30	10.03	100.5	103.2
31	8.928	79.70	130.1
32	7.950	63.21	164.1
33	7.080	50.13	206.9
34	6.305	39.75	260.9
35	5.615	31.52	329.0
36	5.000	25.00	414.8
37	4.453	19.83	523.1
38	3.965	15.72	659.6
39	3.531	12.47	831.8
40	3.143	9.888	1049

Table 2-1. *American Wire Gauge (AWG) system table.*

sistance twice as great for the same given wire length. For example, No. 14 wire is 64.1 mils in diameter and has an area of 4107 cmil with a resistance of 2.53 Ω/1000' at 20°C; No. 17 has a area of 2048 cmil is 45.3 mils in diameter and has a resistance of 5.06Ω/1000.' Yes, there is a No. 17 AWG wire; residential telephone drop wire is parallel No. 17. However, in general, even-numbered wire gauge is used for electrical wiring and the odd-sized wires are used in the construction of coils and the windings of motors, transformers and the like.

By code, the smallest wire that can be used in the United States for electrical (AC) distribution is No. 14.

There are two other measurements for wire and cable, although they are not encountered to any large degree today. These are the B.W.G. (Birmingham Wire Gauge) and the N.B.S.G. (New British Standard Gauge). Occasionally, particularly in some older publications, you might come across a reference to B&S Gauge which stands for Brown & Sharpe gauge; B&S gauge is identical to AWG.

In much of the world the metric system is used for sizing wire and cable. The length will be given in meters and the cross-sectional area will be stated in square millimeters (mm^2).

Questions

1. A 20 ft. length of No. 38 wire has a resistance of 13 ohms. What length of this wire will have a resistance of 8 ohms?

2. A given wire has a diameter of 0.04 in. and a resistance of 2 ohms.

    a. What would the resistance be if its diameter were 0.10 in.?
    b. If the diameter was 0.01 in.?

3. The diameter of No. 14 wire is 0.06408 in. What is its cross-sectional area in circular mils?

4. A bank of lamps is connected with a pair of No. 6 AWG wire to a 230V power source 110 ft. away. If the current to the lamps is 40 amperes, what is the voltage at the lamps' terminals?

5. A 50 ft. length of copper wire has a resistance of 8.2 ohms. What length of the same wire has a resistance of 5 ohms? A resistance of 20 ohms?

(Answers in Appendix A.)

Measurements in DC Circuits

Up to this point we have been discussing the relationships between ohms, volts, amperes and watts strictly from a mathematical viewpoint. The more commonly used instruments to measure these quantities are the multimeter (*Figure 2-48*), galvanometer, the voltmeter, the ammeter, the ohmmeter, and to a lesser degree the Wheatstone bridge, the wattmeter and the megger.

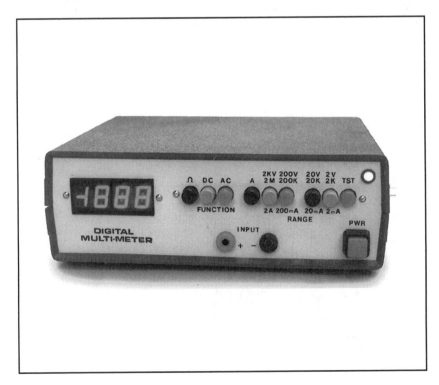

Figure 2-48.
A bench digital multimeter (VOM).

The *galvanometer* is the most elementary electrical measuring device. It ordinarily consists of a coil of several turns of very fine wire suspended between the poles of a permanent horseshoe magnet and held in a neutral position by the torsion of very fine suspension fibers. The suspended coil carries a lightweight needle which stands at the center of a fixed scale when the coil is in its neutral position with respect to the permanent magnet. A very small current through the suspended coil will set up a magnetic field that will tend to align itself with the field of the permanent magnet and consequently cause a deflection of the needle from its neutral position on the fixed scale. A galvanometer may then be considered as simply a sensitive device for detecting direct electrical currents. It is not designed to measure magnitudes of current; but simply just the presence of such currents. The effectiveness of the galvanometer in detecting small value currents is a function of the sensitivity of the meter.

A *voltmeter* (of the revolving coil type) is a galvanometer having an extremely high resistance and with a scale calibrated to read the potential impressed on the terminals (or leads) of the instrument. In actuality, the voltmeter deflection is caused by the very small current flowing through the high-resistance winding. The voltmeter, while it utilizes the principle of a galvanometer, must have its terminals (or leads) designated as positive or negative. The higher the resistance of a voltmeter the better to prevent loading of the circuit being measured.

This is accomplished by using a galvanometer meter movement in series with a high resistance.

An *ammeter* is again an application of a galvanometer with a scale calibrated to measure the value of the current flowing through its windings instead of the electromotive force across its terminals (*Figure 2-49*). The ammeter (of the moving coil type) must be inserted in series with the circuit which it is measuring. Hence, the resistance of the ammeter must be very low so that it does not reflect a change of value in the circuit it is being used to measure.

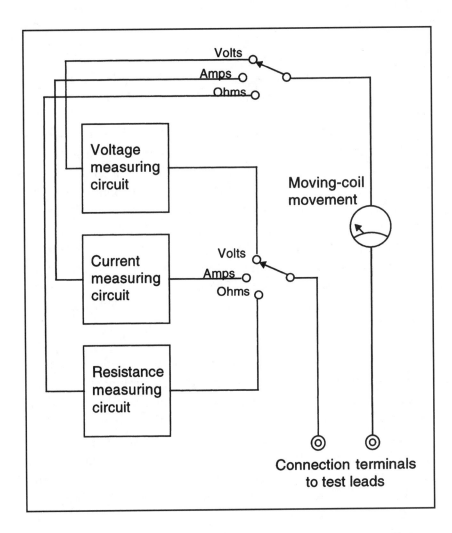

Figure 2-49. *Block diagram showing the functions of an ammeter.*

Because of this very low resistance (on the order of 0.005W) an ammeter must always be inserted and never connected across a difference of potential.

As an illustration of this let's apply Ohm's law to an ammeter whose internal resistance is 0.005W connected across an electromotive force of one volt. Since I = E/R then I = 1/0.005 or I = 200 amperes. This will undoubtedly cause some severe damage to the ammeter.

The circuit depicted in *Figure 2-50* shows a multimeter arranged to measure DC voltages over a range of 2.5 volts to 1000V. Current

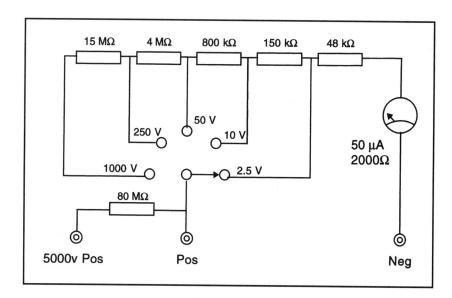

Figure 2-50. *A multimeter arranged to measure DC voltages over a range of 2.5V to 1000V.*

will flow through the meter movement and the voltage scale is set by virtue of the switched series resistors.

If you were to measure the voltage drop across a resistance as discussed earlier, you would connect the + and - leads across the resistor and select the appropriate voltage range.

Figure 2-51 shows the internal circuitry for a multimeter arranged to measure DC current. As noted earlier, the meter must in this case be in series with the circuit being measured. The circuit being measured will be disconnected at one end and the + and - leads will be connected to complete the circuit so that current will flow through the meter to complete the circuit.

A multimeter can also be used to measure values of resistance. In this mode, the circuit being measured must be inactive, i.e., no current flowing.

If you should connect a multimeter in the "ohms" mode across a circuit with current flow you will at best "peg the meter" and, dependent on the current flowing, seriously damage the meter.

127

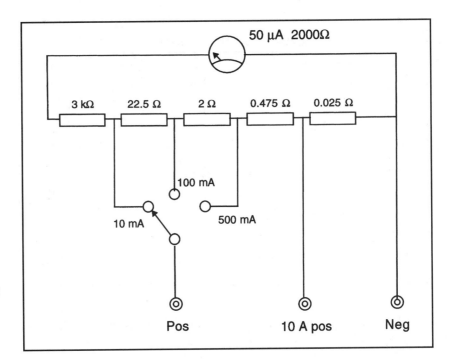

Figure 2-51. *The internal circuitry for a multimeter arranged to measure DC current.*

When used as an ohmmeter, the meter uses voltage from its internal battery. The resistance to be measured is connected in series with the internal battery and the meter movement. The meter will indicate the amount of current flow through the meter movement and the resistance being measured. From Ohm's law we know that current is inversely proportional to resistance. Hence, the scale graduations for resistance are reverse from the scales for voltage and current readings (*Figure 2-52*).

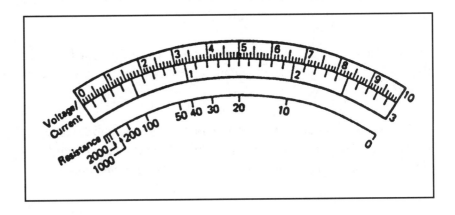

Figure 2-52. *Multimeter scale configurations.*

A "clamp-on" type ammeter which is used to measure the current flowing in a conductor when the instrument is placed around the conductor, is similarly a galvanometer which is used to detect a magnetic field and is calibrated to read the resultant current. A meter of this type is limited to use in measuring AC circuits.

Since an ammeter must have a very low resistance. This is accomplished by placing the galvanometer meter coil in parallel with a low-resistance "shunt" resistor. The shunt then carries nearly all the current, leaving a fraction to pass through the higher-resistance meter.

In practice, both voltmeters and ammeters have various scales to measure very small to very large values of EMF and current. This is accomplished by varying the series resistors in the voltmeter and by varying the value of the shunt resistors in an ammeter

If a voltmeter is used to measure the EMF of a circuit at a point in time and an ammeter measures the current at the same exact time; then, since P = EI the instrument will provide you with a measurement scaled in watts. This is referred to as a wattmeter.

Similarly, since R = E/I then the instrument can be arranged and calibrated to read resistance in ohms. Again, various shunting resistors are arranged so that the instrument can read various scales of magnitude in ohms.

Meggers are instruments which contain a magneto and a speed governing device which produces a fairly high (400 V) EMF. Meggers are generally used for measuring extremely high resistances such as you might find in measuring insulation breakdown values and the like.

A Wheatstone bridge (*Figure 2-53*) is a sensitive instrument used to measure resistance on the principle of a balanced bridge. The principle of a Wheatstone bridge is illustrated as follows: the resistances A and B are the ratio arms of the bridge and the variable

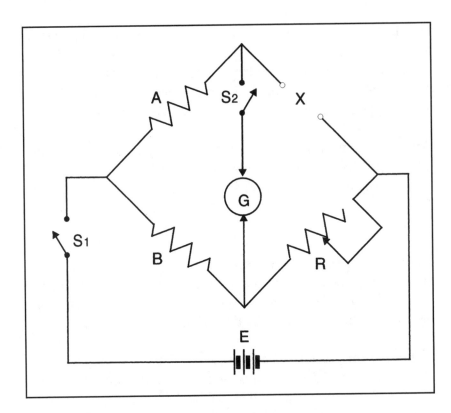

Figure 2-53.
Wheatstone bridge.

resistance R is adjusted so that for any unknown resistance connected at X, the value of R adjusts the galvanometer to a balance usually indicated by a zero point on the scale. Conversely, R could be made to be fixed and the ratio of A and B could be varied to affect a balance.

Wheatstone bridges were used extensively in the telephone industry to measure and test open wire circuits and cable pairs. While not widely used today, the principle of balanced bridge occurs in various versions of modern electronic equipment.

Chapter 3
System Elements

Appreciation is given to the engineering staff at Belden Wire and Cable, Cablec Continental Cables Company, and West Penn Wire Corp for assistance in preparing this section.

- **NICET Work Element 11005, Electronic Components**
- **NICET's Description:** The applicant shall... "identify the characteristics, values (stamped or color- coded) and the circuit functions of resistors, capacitors, and inductors."
- **NICET Work Element 11001, Basic Switched and Connectors**
- **NICET's Description:** The applicant shall... "show ability to connect simple devices, such as microphones, telephone jacks, relays, call-in switches, antenna jacks, timing devices, etc."
- **NICET Work Element 11014, Shop Assembly**
- **NICET's Description:** The applicant shall... "indicate knowledge of the proper means to fabricate assemblies and systems using simple shop tools. Have knowledge of proper means to make terminations, and show an understanding of the rack unit measurement system and signal flow diagrams."
- **NICET Work Element 11012, Basic Wiring**
- **NICET's Description:** The applicant shall... "show knowledge required to pull and install specified wire. Recognize color codes, pull tensions, different wire types, and simple connections."

131

Chapter Contents

ELECTRONIC COMPONENTS
Introduction
Resistors
Capacitors
Inductors
Questions

BASIC SWITCHES AND CONNECTORS
Introduction
The Evolution of Switches
Types of Switches
Relays
Call-In Switches
Audio Connectors
RF and Antenna Connectors
Telephone Jacks
Questions

SHOP ASSEMBLY
Introduction
Shop Assembly of Loudspeakers
Electronic Equipment Rack and Panel Assembly
Fabricating the Equipment Rack
Wiring the Equipment Rack
Equipment Terminations
Soldering Techniques
Tools

BASIC WIRING
Introduction
Types of Wire and Cable
Cable Installation Procedures
Physical Installation of Wire and Cable
Equipment

Conduit Fill
Ambient Temperature
Cable Connections

ELECTRONIC COMPONENTS

Introduction

Thus far in our studies, we have examined the theoretical effects which circuit components exhibit on the voltage and currents of DC circuits. We have learned that specific lengths and gauges of wire and cable have specific values of resistance that can be readily calculated. In addition, we will be examining how inductors and capacitors cause reactance in a circuit and how these properties interact to form resonant circuits.

If you examine a schematic of an electronic circuit or device, it quickly becomes apparent that in addition to the active elements (transistors, diodes and integrated circuit chips), the circuit contains a multitude of resistors, capacitors and inductors. These electronic components are utilized to vary and modify the electronic circuits so that the circuit will operate as the designer intended.

A major part of an electronic technician's vocation involves troubleshooting electronic circuits. Electronic circuits do malfunction. And, when they do malfunction, it falls on the technician to determine why the component failed and then correct the fundamental problem. It is not sufficient to merely replace the failed component. The technician must determine the root cause of the failure and also correct it. Otherwise, the failure becomes self-perpetuating.

When an engineer designs an electronic circuit for a particular device, he or she chooses specific values of resistance, capacitance and inductance to ensure that the circuit performs in a manner for which it was intended. Hence, the circuit may contain a multitude of varying values of these components.

A schematic will not show the physical properties of the component. For example, in the case of a resistor, it will show a representation which we interpret as a resistor. The particular value of the resistor will either be shown adjacent to the representative symbol or in a table of values on the schematic. It is not uncommon to label components as R1, C3, Q106, L12, etc. The technician working on such a circuit may find it necessary to refer to the material list to find the actual value of a specific component.

Electronic Components, Symbolic Representations
From our studies to date, we have been exposed to the conventional symbols for various electronic components. (See *Table 3-1*)

There are also various symbols that depict (schematically) functions such as connecting plugs and patching points, transformers, attenuators, pads, indicating lamps and LEDs, and all of the other components that are used to construct an electronic circuit. We will cover these additional circuit components when we study schematic diagrams and circuit analysis later in this book.

	SCHEMATIC SYMBOL	PROPERTY	UNIT SYMBOL
1.	—/\/\/\—	Resistance	ohms, R, Ω
2.	—||— —||⁻	Capacitance	farads, C
3.	‒0000‒ ‒0000‒	Inductance	henrys, L
4.	⁺—◁|—₋	Rectifier	Diode, CR
5.	⊗ ⊗	Transistor	Q101, Q708, etc
6.	—⁺|||⁻—	Battery	EMF, V , voltage

Table 3-1. Symbols for various electronic components.

Resistors

Resistors are available in values ranging from fractions of an ohm to millions of ohms. They are employed extensively in electronic circuits. Although the value of resistance may be an integral part of an integrated circuit (and thus not be physically identified as a separate component), the student of electronic circuits should be aware that the property of resistance is still influencing the operation of the particular circuit.

Resistors come in a variety of construction forms. Some of the more typical mechanical constructions are shown in *Figure 3-1*.

Figure 3-1 shows three different styles of resistors we commonly find in our line of work. The first resistors are fixed resistors. Although both are tubular in design, one is made of a carbon

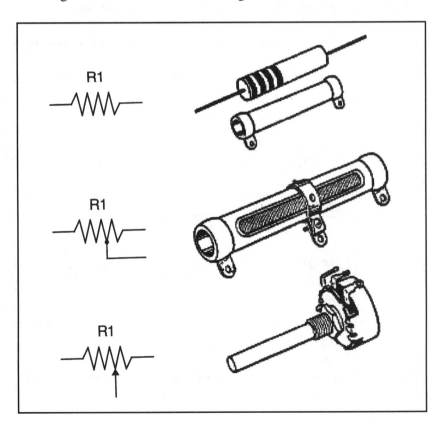

Figure 3-1. *Typical resistor mechanical construction forms.*

composite material and the other is a wire-wound design. You should know that all wire-wound resistors will also have some inductance, like a coil, due to its coil-type construction. Wire-wound resistors are usually used in high-power applications involving DC or fixed-frequency AC circuits. You should not use a wire-wound resistor for a dummy load with an audio amplifier, as the inductive component of the resistor will change the resistor's impedance depending on audio signal frequency content. Fixed resistors are also made using a metal film material that provides excellent precision and stability for resistive values.

The second type of resistor (shown in *Figure 3-1*) is a fixed-variable wire-wound resistor. The slider can be moved to vary the resistance between the slider and one end of the resistor, then tightened down to maintain a particular resistance. A variation of this resistor allows the slider or wiper to be continuously variable, thus creating a resistor known as a *rheostat*.

The third resistor shown is a variable resistor, also known as a potentiometer, or "pot." Most commonly used for volume controls, the pot will usually be made with a carbon composite, although some will use a wire-wound construction.

The resistance value of a discrete resistor (in ohms) is generally referenced on the schematic drawing or on a parts list, which is a part of the schematic. In addition, the resistance value of an individual resistor is generally shown on the component itself. In the case of physically large carbon resistors, the value may be stamped on the component. This is also the case with physically large variable resistors, where the resistive value will generally be shown on the body of the device. In the case of smaller composition (carbon) tubular resistors, the value of the resistance is indicated by a color coding on the body of the resistor.

In the case of carbon resistors, it is important that we learn, *and remember*, the color-code which will allow us to determine the value of any specific, discrete resistor.

An example of a typical composition-type resistor is shown in *Figure 3-2*. It will be noted that the resistor component has "rings" of colored bands that can be "read" to determine the value of the resistor by referring to the "resistor color-code" shown:

RESISTOR COLOR-CODE

Color Band	Significant Figure	Multiplier Value	Tolerance %
Black	0	1	
Brown	1	10	±1
Red	2	100	±2
Orange	3	1,000	±3
Yellow	4	10,000	±4
Green	5	100,000	±5
Blue	6	1,000,000	±6
Violet	7	10,000,000	±7
Grey	8	100,000,000	±8
White	9	1,000,000,000	±9
Gold		0.1	±5
Silver		0.01	±10
No Color			±20

In looking over this chart, you find how to determine the resistor's value using the first three colored bands. The fourth band, if there is one, indicates the percentage of tolerance that the particular resistor exhibits; that is, how much above or below the marked

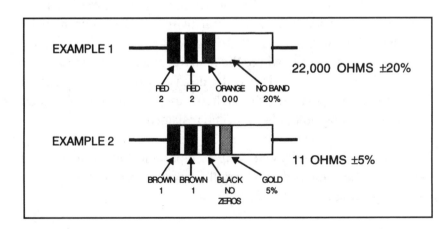

Figure 3-2. *Typical composition-type resistor.*

138

value can the resistor actually measure, under normal operating temperatures, and still be consider the marked value. A resistor with no fourth band is assumed to have a tolerance of plus or minus (±) 20% of the marked value. The tolerance of a resistor can be an essential factor in circuits, like filters, that require precise values of resistance.

While it might at first thought seem strange to add resistance to a circuit and thus create a change to the current flow, it will be seen that this is a legitimate method to make the circuit "behave" in the manner that the designer intended. For example, we will learn in the resonance section that the characteristics of a series or parallel resonant circuit can be modified by introducing a specific value of resistance to the LC network.

In addition to exhibiting a value of resistance (within the tolerance percentage indicated), a resistor will also be "rated" as to its power rating. That is, it will be defined as being able to dissipate some value of power.

As we learned in an earlier chapter, we can calculate the power being dissipated in the resistor by using Ohm's law. To calculate the current through the resistor, we divide the voltage across the resistor by its resistance. Then by multiplying that current times the voltage, we find the power:

$$I_{amps} = E_{voltage}/R_{ohms}, \text{ then } P_{watts} = E \times I$$

The direct formula to calculate power, given only resistance and voltage, is:

$$P = E^2/R$$

If the resistor's power rating is exceeded, there is a good likelihood that the resistor will self-destruct. Resistors should be selected with a power rating large enough to prevent significant heating, which can damage the resistor as well as surrounding components and circuit boards. It is not unheard of to have undervalued resistors burst into flame.

Some large power resistors will be stamped with the actual power rating. In most cases, the only indication of the power rating is the physical size of the resistor.

Always calculate the power to be dissipated when selecting resistors for your circuit design.

Capacitors

First of all, let's correct a misconception. Capacitors are often referred to as "condensers." This is an antiquated term that has no real use in the modern electronic technician's vocabulary, but you will run into it in some electronic texts and when dealing with some microphones. You should be aware that the word "condenser" is referring to a capacitor or a design based on capacitor principles.

As we learned earlier, a capacitor is composed of two metallic or conductive surfaces separated by some insulating material, which is referred to as the "dielectric" of the capacitor. The capacitor is the only device other than a battery that can store electrical energy. The "capacitance" of a capacitor is its ability to store an electric charge. The capacitor's capacitance is measured in units known as farads.

Capacitors come in a wide variety of shapes and forms. Several of the various types of physical configurations of capacitors are shown in *Figure 3-3*.

Fixed capacitors may take the form of various constructions. They may be constructed by using dielectrics of paper, mica, Mylar, air and other materials.

One special type of capacitor is the *electrolytic* (frequently used in power supply filtering circuits). The electrolytic capacitor is used

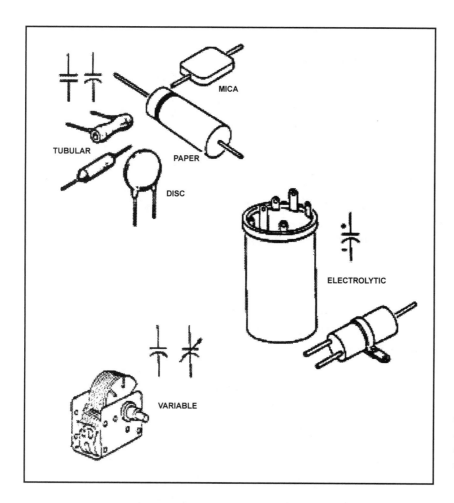

Figure 3-3. *Various types of physical configurations of capacitors.*

where very large values of capacitance are required. (An electrolytic capacitor may have a value of several thousands of microfarads of capacitance.) Electrolytic capacitors are *polarized*; that is, one plate of the capacitor must be operated at a positive DC potential above the other plate. Consequently, when dealing with electrolytic capacitors, you must observe the polarity of the component. Polarized electrolytic capacitors have a special symbol, as shown in *Figure 3-3*.

Like a resistor's maximum power rating, capacitors have a breakdown voltage rating. A capacitor may have a rating of 50 WVDC (working voltage DC). This would indicate the device will

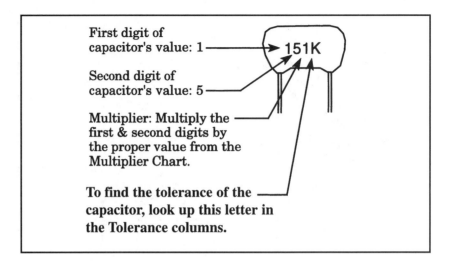

First digit of
capacitor's value: 1

Second digit of
capacitor's value: 5

Multiplier: Multiply the
first & second digits by
the proper value from the
Multiplier Chart.

**To find the tolerance of the
capacitor, look up this letter in
the Tolerance columns.**

*Figure 3-4. Most
capacitors have their
values stamped
directly on them.*

tolerate a maximum of 50 volts impressed on its terminals. A voltage
exceeding the rated value can cause the capacitor to fail due to the
breakdown of the dielectric and the arcing of voltage between
capacitor plates.

Inasmuch as capacitors come in a wide range of constructions, the
method of physically defining the value of capacitance of an
individual capacitor varies rather widely.

Most capacitors are "stamped," or otherwise identified as to their
value, directly on the body of the unit (*Figure 3-4*). These capacitors
are usually marked with values in microfarads (mfd or μF) or
picofarads (pF or mmfd). The microfarad represents 10^{-6} farads,
while the picofarad represents 10^{-12} farads.

A coding system has been established to allow marking on small
capacitor bodies. The following diagram and chart indicate how
these marking are read.

Multiplier		Tolerance of Capacitor		
For the Number:	*Multiply by:*	*10pF*	*Letter or Less*	*Over 10pF*
0	1	±0.1 pF	B	
1	10	±0.25 pF	C	

2	100	±0.5 pF	D	
3	1000	±1.0 pF	F	±1%
4	10,000	±2.0 pF	G	±2%
5	100,000		H	±3%
			J	±5%
8	0.01		K	±10%
9	0.1		M	±20%

Examples:

151K = 15 x 10 = 150 pF ±10%

759 = 75 x 0.1 = 7.5 pF

Note: The letter "R" may be used at times to signify a decimal point; as in:

2R2 = 2.2 (pF or μF)
pF = picofarads; μF = microfarads

Some paper, ceramic or mica capacitors, particularly military-rated ones, will use a color-coded marking system similar to the one used for resistors.

As can be seen from *Figure 3-5*, the mica-type capacitor, instead of being color-coded by bands of color, is identified by colored dots on the body of the component.

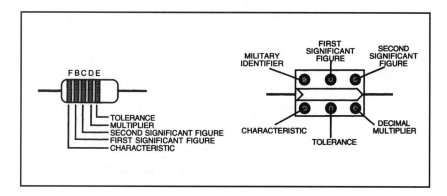

Figure 3-5. The mica-type capacitor is identified by colored dots (see Capacitor Color Code).

CAPACITOR COLOR CODE

Color	1st Sig. Fig. (B)	2nd Sig. Fig. (C)	Mult. (D)	Tolerance (E)	Characteristic (F)
Black	0	0		±20%	F
Brown	1	1	10		
Red	2	2	100	±2%	B
Orange	3	3	1,000		C
Yellow	4	4	10,000		D
Green	5	5	100,000		E
Blue	6	6	1M		
Violet	7	7	10M		
Grey	8	8	100M		
White	9	9	1000M		
Gold			0.1	±5	
Silver			0.01	±10	

Note: Capacitance in picofarads

Capacitors are widely employed as variable units for tuning resonant circuits. As we shall see, it is much easier (mechanically) to vary the capacitance of a circuit than it is to vary the inductance of the circuit.

When we vary the position of the capacitors' plates with respect to each other, we affect a change in the relative capacity of the component. One of the most common examples of this type of component is the "tuning capacitor" in your AM/FM radio receiver. If you examine the tuner of your radio set, you will see a "ganged" capacitor (*Figure 3-6*), which consists of two metal sheets which are interleaved and mounted so that one set of plates can rotate into or out of mesh with the other set of plates. By changing the ratio of the area of plate set A with plate set B, we effectively vary the capacitance of the assembly.

Various combinations of rotor elements and stator elements (the rotating plates are referred to as the rotor elements, and the static or

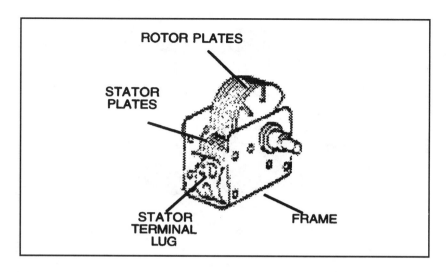

Figure 3-6. A ganged capacitor.

fixed plates are called the stator pieces) may be used to vary the capacitance of different circuits within the assembly. This type of construction is frequently employed in radio and TV receivers where it is desired to change the tuning of the receiver. Thus, it might be said that "ganged capacitors" are used to "track" two or more circuit elements and vary their respective capacitances to provide the desired effect.

Capacitors are used in electronic circuits to modify the behavior of currents and voltages in any given circuit. Because the dielectric is an insulator, a capacitor does not permit a direct current to flow through it; but its continuous charging and discharging action permits alternating current (AC) to pass. One of the main uses of capacitors is to block DC but pass AC from one amplifier to another. Also, because of the capacitor's ability to store energy and its opposition to any change of voltage across its terminals, capacitors are used in power supply circuits to filter or smooth out the DC voltage being created from an AC source.

Capacitors are used for many different purposes in electronic circuits. With the vast variety of styles and types, the capacitor is one of the most widely-used components.

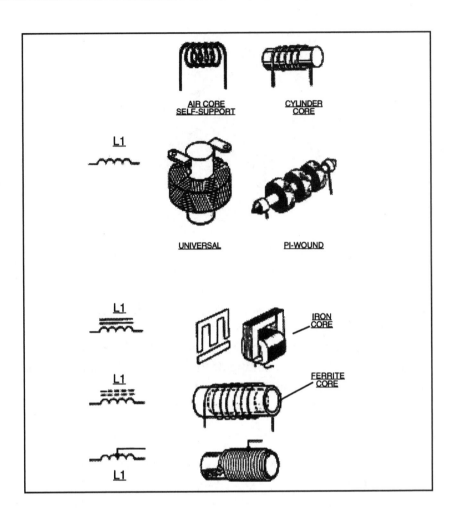

Figure 3-7. Various physical forms of inductors.

Inductors

Inductors, also known as *chokes* or *coils*, exist in several different forms of construction in electrical and electronic circuits. Their purpose is to introduce some value of inductance into the circuit in which they are employed. Inductance is a property of an electrical circuit in which voltage (EMF) is induced by a change in current. The symbol for inductance is L. It is measured in units of henrys, named for Joseph Henry (1797-1878).

The various physical forms of inductors are displayed in *Figure 3-7*, along with the schematic representation of their respective forms.

We learned earlier that in an electromagnetic coil, the strength of the magnetic field is proportional to the rate of change of the electrical current flowing in the circuit. Hence, we were able to determine that an "air-core" inductor would have a lower degree of magnetic inductance than the same coil wound on an iron-core coil.

An inductor can be made to have more or less inductance by varying the number of turns or by employing different materials for the core.

Inductors may also be constructed to have a variable inductance. This principle is often employed in radio transmitter and receiver sets, where it is necessary to vary the inductance in an intermediate-tuned circuit of the device. Again, an examination of your radio will show you a number of "IF" intermediate frequency coils, which are basically series inductors used to "tune" your receiver's "IF" during manufacture. The construction of a variable inductor and its schematic representation are shown in *Figure 3-8*.

The specific value of an individual inductor is generally shown by a stamp or mark on the component. If not, you will have to refer to the schematic "bill of material" to determine the value of the inductor.

Inductance in a circuit is generally employed in power supply filtering networks, crossover networks and resonant circuits. In high-level crossover networks for loudspeakers, inductors are used

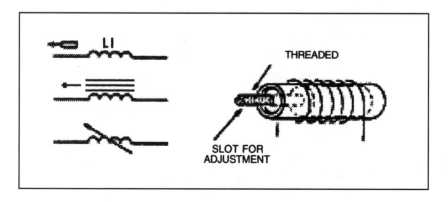

Figure 3-8. A variable inductor.

147

extensively to "separate" low-frequency audio signals from higher-frequency audio signals.

An inductor should be considered as a "frequency dependent" component that is utilized to "tailor" the frequency characteristics of the circuit in which it is employed.

Questions

1. Indicate schematically the symbol for:

    a. A variable resistor.
    b. An electrolytic capacitor.

2. A carbon resistor is color-coded green, black, brown, silver. What is its resistance and tolerance?

3. What is the unit of measurement for a capacitor?

4. A ceramic capacitor is color-coded violet, gray, black. What is its value and its tolerance?

5. What is the letter symbol for inductance?

6. Give three examples of how the property of inductance can be used in electronic circuits.

7. A value of capacitance is wired in parallel with a value of inductance:

    a. What does this circuit constitute?
    b. Where might such a circuit be employed?

8. What is the purpose of using electrolytic capacitors in a power supply circuit?

(Answers in Appendix A.)

BASIC SWITCHES AND CONNECTORS

Introduction

A switch is very simply a device that opens or closes a circuit, or reroutes a signal from one path to another.

We learned in basic DC theory that air is a good insulator. Hence, when an electrical circuit is placed in the "open" position, the air forms a very high resistance and an electrical current ceases to flow.

We use switches almost constantly:

1. When your alarm goes off in the morning, it either switches on the radio or the alarm to tell you it's time to get up.

2. You grope around and "switch" on the light in your bedroom.

3. You stumble into the kitchen and "switch" on the coffee pot.

4. You "tune" or "switch" your kitchen radio to your favorite morning radio show.

5. You place two pieces of bread in the toaster, depress the lever and "switch" on your toaster.

6. You go out and start your car by "switching" on the ignition. (Hopefully, it starts.)

7. On your way to work, traffic lights "switch" on and off from green to yellow to red (with an occasional arrow).

The Evolution of Switches

From the earliest experiments with electricity, it was recognized that some means must be made available to shut off the flow of

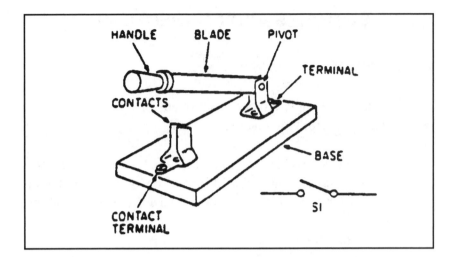

Figure 3-9.
A single pole-single throw (knife) switch.

current to a circuit. Obviously, the first way to accomplish this was to disconnect one or both leads from the battery to the circuit. However, physically disconnecting the wires from the battery terminals was not the most efficient manner to accomplish this task. Hence, early on, it was recognized that if you opened (switched one of the legs of the circuit into a resistance of infinite resistance) the circuit, you would cause a cessation of current flow.

One of the earliest and most rudimentary type of switch is shown in *Figure 3-9.* (These were generally referred to as "knife switches.") The one shown was referred to as a single pole-single throw switch (SPST). Obviously, it could only "open" or "close" one leg of a circuit.

Figure 3-10.
DPST switch.

While the illustration depicts a physical construction seldom encountered today, the basic principle of a SPST is still in vogue in the basic house lighting system switch.

In some electrical circuits, particularly those employing polarity-sensitive circuits, it became necessary to introduce a double pole-single throw switch (DPST). This type of switching circuit is also practical when you want to turn two circuits "on" at the same instant.

An illustration of this type of switch and its schematic diagram is shown in *Figure 3-10*.

In this type of configuration, both legs of the circuit have corresponding connecting terminals for both sides of the associated legs.

This type of switching circuit, while physically different, is still widely used to switch-balanced signal level circuits.

In some instances, it might be desirable to cause a cessation of current flow in either the source or the terminating circuit. Or it might be desirable to route the signal path from one circuit to an entirely different path.

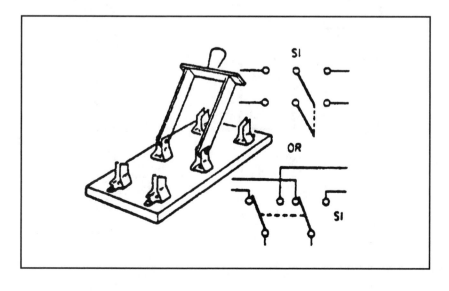

Figure 3-11.
DPDT switch.

To accomplish this type of circuit transfer, the double pole-double throw (DPDT) switch was introduced (see *Figure 3-11*).

If the signal source is connected across the terminals of the knife switch, then the signal can be routed to either circuit "A," corresponding to the upper set of terminals, or circuit "B," corresponding to the lower set of terminals.

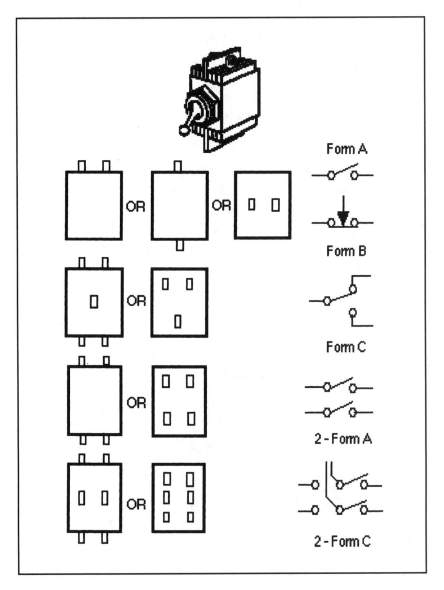

Figure 3-12. The four most common types of toggle switches.

Every switch operates on the concepts shown, although the number of poles and throws may vary widely.

Types of Switches

Knife switches have been largely replaced with more modern types of components. These are commonly referred to as *toggle* switches. These are enclosed, bakelite or plastic composition shells with enclosed terminal connections.

The four most common types of toggle switch schematics are illustrated in *Figure 3-12* along with their schematic diagrams. As illustrated, the physical connections may be on the rear of the switch, at either end of the switch, or at either the top or the bottom of the switch.

From these rudimentary concepts, we have developed extensively complex switching networks. These will be the subject of further discussions in this series relative to advanced switching networks.

Switches sometimes have special contact arrangements, including the momentary push-button, both normally-open and normally-closed, make-before-break and its counterpart, the break-before-make contact. Almost any contact arrangement you can think of can be built into a switch.

Another special form of a toggle switch is the so-called *slide switch*. This type of switch functions like a toggle switch but operates by means of moving a handle or paddle from left to right, or up and down, depending on how it is physically mounted. A diagram of the mechanical construction and the representative schematic diagrams for this type of device is shown in *Figure 3-13*.

As shown, the slide switch can take on the same type of circuit configurations as a toggle switch; i.e., SPST, SPDT, DPST, etc. They, like toggle switches, can be center off, left on or right on. Both toggle switches and slide switches come in a variety of sizes

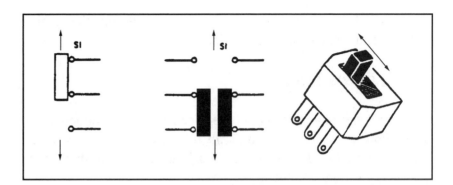

Figure 3-13.
Slide switch.

and have switching contact capacities from very minute to large current handling. Slide switches are frequently employed in circuits that require minimum operation. They are often found in circuits that could be considered "set-and-leave" adjustments. Inasmuch as they are mechanical devices that depend on the movement of a set of leaves, frequent operation could result in mechanical failure.

Another special application of the slide switch is the *dip switch*.

Dip switches are usually subminiature, ganged slide switches. An example of a typical dip switch is illustrated in *Figure 3-14*. These are often employed in circuits which require multiple combinations to affect the particular circuit operating characteristics. Selection of filter operating parameters comes to mind as the type of circuit where this switching device might be employed.

When it is necessary to switch a signal source to a number of different circuits, the *rotary switch* is commonly employed. These

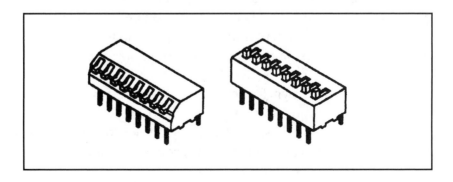

Figure 3-14.
Dip switch.

types of switches are being rapidly replaced with digital signal processing (DSP) networks; however, they still exist in a multitude of circuits which you may encounter.

A rotary switch may employ one or more ganged sections. Each switch section may be wired to create varying conditions in the circuit associated with each "gang," or wafer.

A depiction of the mechanical construction of a simple "one gang" and a multiple "three gang"-type rotary switch is illustrated in *Figure 3-15*. Note this switch is an 11-throw switch with each ganged section an additional pole.

Intercommunications and radio frequency circuits make extensive use of rotary switches inasmuch as the switched characteristics of circuit "A" may affect the desired effects on circuit "B," "C," etc.

A very simple example where a rotary switch might be employed is where the outputs of multiple amplifiers need to be routed to a

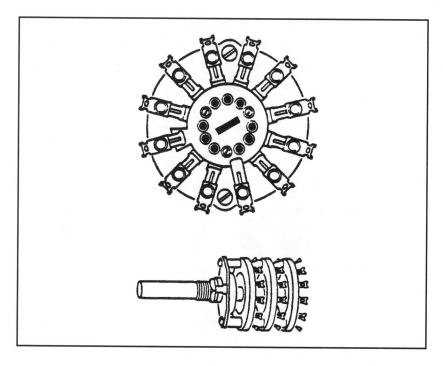

Figure 3-15. One gang and three gang-type rotary switch.

monitor panel. Each amplifier's constant output voltage would be connected to the rotary switch input (contact) connections, and the monitor speaker or level indicator would be connected to the rotary switch wiper contacts. By moving the rotary switch from position one to position eleven, the output integrity of each amplifier's output can be audibly or visually verified.

Relays

A relay is very simply a remote-controlled switch. Relays can be either electromagnetic or electronic. In this section we will confine our discussions to the electromagnetic type of relay, and leave the electronic switching circuits for a later section. As we learned earlier, when an electrical current is made to pass through a coil winding, a corresponding magnetic field is established surrounding the coil. Consequently, when a current is applied to a relay coil, the coil becomes magnetized. As shown in *Figure 3-16*, an illustration of a simple relay, when the coil is energized, the relay armature is attracted to the core.

The contacts associated with the movement of the armature are then caused to mechanically change their state. Also, as shown, the contacts may be normally open (NO) or normally closed (NC). The *normal* condition refers to the state of the contacts when the coil is de-energized.

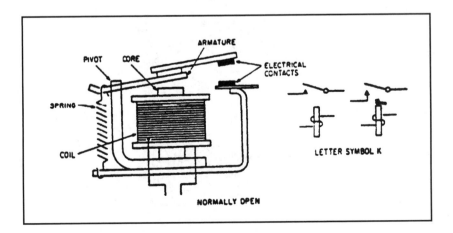

Figure 3-16. Simple, open frame type of relay.

Relays are constructed in a variety of sizes, types, current-switching capacities, coil types and contact arrangements. The simple relay depicted in *Figure 3-16* is of the open-frame type.

The terms "relay" and "solenoid" are frequently used interchangeably; however, a solenoid is more generally associated with circuits where the contacts are switching AC power or have some required mechanical action.

As mentioned, relays can have multiple contact arrangements within a common structure. The drawing in *Figure 3-17* shows, schematically, a simple SPDT (form C) relay contact arrangement and a more complex multiple-contact arrangement. Other generic contact arrangements are the form A, which is an SPST normally-open switch and the form B, which is an SPST normally-closed switch. Almost all other contact arrangements are a compilation of these three basic forms.

Mechanically, the contacts, the swinger and the coil leads are brought out to some form of terminal connection. The connector may be of

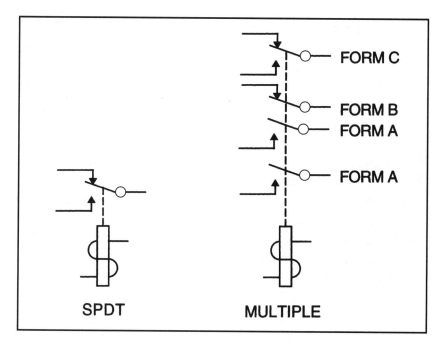

Figure 3-17. A simple SPDT relay contact arrangement and a more complex multiple-contact arrangement.

the solder-type terminal, or the connections may appear on an octal-socket type (plug-in type) connector.

It is considered good engineering practice to install a kickback or shunting diode across the relay coil, to suppress the back EMF generated by the collapsing magnetic field of the coil when it is de-energized. While normal relays are not polarity sensitive, proper polarity must be observed after installation of the diode, or a short circuit will occur.

Call-In Switches

Most forms of intercommunication circuits (intercoms) use some form of call-in device. This allows a user at one end of the circuit to notify the selected party at the other end that the user wants to talk to the called party. This type of circuit is commonly employed in school intercom systems, where a call-in switch is located in the various classrooms. By operating the call-in switch, the classroom instructor can notify the master station in the office that the instructor wishes to talk to the office.

Another form of call-in is the typical patient station in a *nurse-call* system. Nurse-call systems are somewhat more complex than school intercom systems, and generally involve turning on signalling systems in addition to advising the master that the patient needs to communicate with the nurse station. We will cover nurse-call systems in greater detail in Level II.

The voice call-in is almost obsolete. At one time most call-ins were accomplished this way. The voice call-in is in actuality a DPDT switch. This switch, mounted on a wall plate, was a momentarily operated switch that switched the room intercom speaker from the home-run audio pair to a circulating voice call-in pair. The classroom instructor would push the button and announce their location into the speaker.

The call would be heard at the office and the office person would flip the switch associated with the announced room and facilitate a two-way conversation by operating the talk-listen switch. The room speaker serves as both the microphone and speaker in the intercom system.

Most modern intercoms use a form of annunciator call-in. The classroom call-in switch is normally a momentary SPST push-button, unless privacy features are included. Then the switch might be a rocker-type switch with a momentary call position, a normal mode position, and a privacy mode position.

Wiring schemes for annunciator call-ins vary between manufacturers and system type. However, the basic premise is the momentary switching of a ground (usually the shield of the home-run cable) to an annunciator wire, thus providing a path that triggers a light and some type of audible indicator at the office.

Audio Connectors

XLR

There are probably as many different kinds of audio connectors as there are pages in this book. But the good news is, as a Level I technician, there are only three types to know inside and out.

The first connector is commonly referred to as an *XLR connector*. The designation changes by manufacturer, but the connectors themselves are interchangeable. The 3-pin XLR-type connector is the connector of choice for all balanced microphone level applications. The connector is also used for balanced line-level applications. Even a few manufacturers will use them for speaker level applications.

Ever since Cannon introduced the XLR, a debate has raged over which of the three pins should be considered positive (high) and which should be called negative (low). It really doesn't make any difference, as long as the convention is held consistent, but a standard was missing.

Finally, after 30 years of debate, the Audio Engineering Society (AES) issued a standard (S14-1992\.ANSI s.4.48.1991) that adopted the convention that pin 1 is ground (shield), pin 2 is hot (positive) and pin 3 is low (negative). This standard is to be followed for all applications using the connector.

The ground lug provided in the XLR is used to connect the connector shell to the shield and, in turn, to ground when the plug is inserted into a receptacle whose shell is grounded. It is left unconnected unless the manufacturer specifically requests a connection to the shell. If connected, it can completely destroy an isolated AC grounded system.

The XLR connectors come in both male and female configurations, and also in both cord mount and plate mount styles. In every case, the technician should carefully prepare the wires and the connectors by pretinning. With cables properly prepared, exposing the proper amount of tinned conductor (1/8"- 1/4"), the soldering iron should be applied to the connector pin. As the solder begins to flow, the conductor should be inserted. The soldering iron should be removed as soon as the solder flows into the wire, to avoid melting the conductor insulation. A proper connection, which has a shiny look to it, shall hold the conductor firmly.

All conductors are attached in the same manner. The shield or drain wire should be sleeved with an insulating tube before attachment to one of the other conductors or the shell, to prevent accidental shorting. Be sure to use the correct type of connector for the cable size being used. The strain relief of the connector must be able to be fully engaged to prevent damage to the conductors when the cable is stressed.

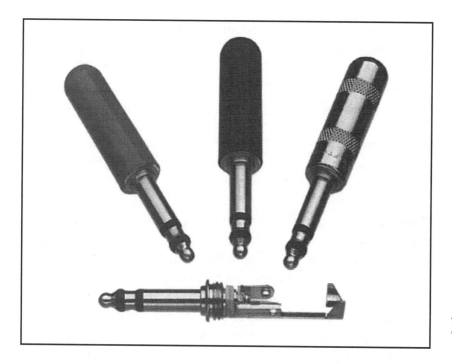

Figure 3-18. *Two-circuit phone jacks.*

Phone

The phone connector, also know as the 1/4" connector, comes in several styles and is the workhorse of our industry. It is used for microphone inputs, line level inputs and outputs, speaker level inputs and outputs, and a myriad of other applications.

The connector was a mainstay of the telephone industry in its infancy; hence the name *phone jack*. The phone jack or plug comes in two varieties: a single-circuit connector, and the two-circuit connector (see *Figure 3-18*).

The very end of the connector is called the TIP. If it is a two-circuit device, the insulated connecting surface right after the TIP is called the RING, and the rest of the connector shaft is called the sleeve. In telephony, the TIP and RING were used to connect the audio path in early cord-board exchanges. The term is still used to identify the connection of a telephone circuit.

161

If the connector is a single-circuit device, the RING is left off the plug, and no corresponding connection is available in the jack. Single-circuit 1/4" phone plugs are used mainly for high-level unbalanced audio connections and for speaker-level connections between an amplifier and portable loudspeaker units. The single-circuit 1/4" phone plug is also sometimes used for unbalanced microphone connections. These are used mostly for high-impedance microphones, but also for low-impedance microphones in some semipro gear.

The two-circuit 1/4" phone plug used to be the main connector for balanced microphone applications. However, most professional and semipro gear is changing to the XLR. It is still used in home equipment for microphones, but its main use now is for balanced line-level connections between the mixer and outboard equipment. Special configurations will use the two-circuit connector as a patching plug, using the TIP to send a signal out and the RING to receive a signal. Both signals are unbalanced and use the sleeve as a common ground.

This type of configuration is used in stereo headphone circuits where the hot lead of one channel (say the left) is put on the TIP, and the other (right channel) on the RING, with the sleeve as the common connection.

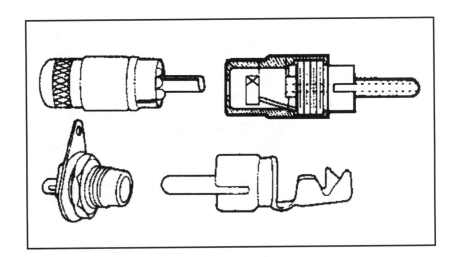

Figure 3-19. *Phono (RCA) connector.*

The 1/4" phone connector is made by many manufacturers in many different styles for different uses. You should be aware of the manufacturer's specifications as to how the connection to the connector should be made. Some are solder, some are crimp, and some use screw terminals. The type you use will depend on your application.

The phone connector also comes in several smaller versions. Examples are the 1/8" tiny plug and the even smaller microplug.

Phono
The phono, also know as the RCA, was developed by RCA and used mainly for connections in phonographic equipment. The connector is used almost exclusively for high-impedance, unbalanced, line-level applications.

The connector (see *Figure 3-19*) has a center pin that serves as the high or positive connection, and a surrounding sleeve which is the ground or low connection. This type of connector is usually soldered onto the cable, but there are some crimp types now available. The panel version of the jack usually has the sleeve connected to the chassis by virtue of its construction. However, there are some models that are built onto insulators, so the sleeve connection can be independent of the chassis.

This low-cost connector has become the mainstay of the consumer audio industry, and is also used in some semipro and pro gear.

Again, proper preparation of your cable prior to installation of the phono connector will go a long way in making the process an easy one.

RF and Antenna Connectors

There are a great variety of connectors available to use with antennas, video and RF signals. As a Level I technician, you need to know the basics.

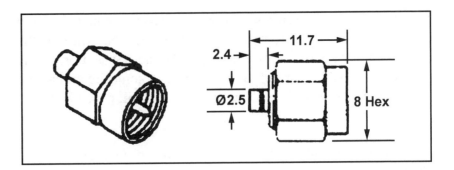

Figure 3-20.
"F" connector.

"F" Connector

One of the simplest and most often used connectors is the "F" connector. In its basic form, the "F" connector uses the solid inner conductor of a coaxial cable as its hot or signal connection. The connector itself is crimped to the shield and outer jacket to serve as the ground or common connection. The connector is threaded, and screws onto the mating "F" barrel connector (see *Figure 3-20*).

Because this connector is used with RF frequencies, the connector, like the coaxial cable on which it is attached, is made with a characteristic impedance. Almost all "F" connectors are 75W devices made to work with RG-59, RG-6 or RG-11 type cables.

These connectors are most commonly used to interconnect components in an RF distribution system. They are used for components like splitters, antennas, amplifiers, filters and television sets.

Follow manufacturer instructions on cable preparation prior to the crimping of the "F" connector. Screw-on connectors are available as well, but are not as commonly used.

Figure 3-21.
PL connector preparation.

PL Connector

The PL-259 or PL-256 (also known as the UHF connector) has been around for many years. They are much larger than the "F" connector and provide a much more substantial connection.

These connectors come in both 50W (PL-259) and 75W (PL-256) versions (some other special impedances can also be found). They used to all be solder-on devices, but new crimp-on styles are being more commonly used nowadays (see *Figure 3-21*).

The main uses for this connector are in wireless microphone system antennas, professional RF equipment, and in some consumer-grade CB-type equipment. The impedance of the connector will be dictated by the cable it is used on and the impedances of the end devices.

Again, like the "F" connector, closely follow the manufacturer's instructions for cable preparation and crimping. Again, screw-on connectors are available as well, but are not as commonly used as the crimp-on and solder type.

BNC Connector

The last type of RF connector we are going to look at is the BNC connector. This connector is also available in 50Ω and 75Ω models (with other special impedances also available). For "precision" systems, use 50Ω for RF, 75Ω for video, and 75Ω for digital audio. Most BNCs in use are 50Ω.

The connector, smaller than the PL, has a twist-and-lock feature that makes it very desirable in professional RF applications (see *Figure 3-22*). The center conductor of the coaxial cable is soldered or crimped to a center pin, which is inserted into the body of the BNC connector. The body of the connector is crimped to the shield and outer jacket of the coax cable, forming a solid stress-reducing connection. This connector offers precise impedance matching and very low loss connections.

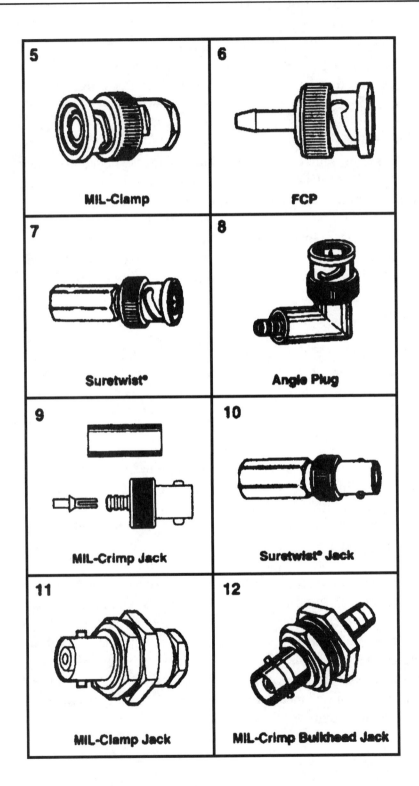

Figure 3-22.
BNC connectors.

This connector is usually found in professional and semiprofessional video equipment. They are also being used by some manufacturers for wireless microphone receiver antenna connections.

Specialized tooling is needed to correctly mate the BNC connector to the cable. Again, the technician must carefully follow the manufacturer's instructions for cable stripping and preparation to ensure a good installation.

Telephone Jacks

Although, as audio technicians, we don't commonly deal with telephone systems or their interconnection, we do occasionally need to interface with telephone equipment, as in the case of a paging system.

The jack or connector most commonly provided in telephony is a 6-pin device known as a RJ-11, or a RJ-14. RJ-11 and RJ-14 are USOC (Universal Service Operating Code) designations, which describe how the jack is connected. In most cases, the connector is identical and the code is only indicating whether one or two telephone lines are connected to it.

Figure 3-23 shows an RJ-14 jack. The middle two pins of the jack carry the TIP and ring of the first telephone line (designated T_1/R_1), and the two outside pins carry the second telephone line (T_2/R_2). If

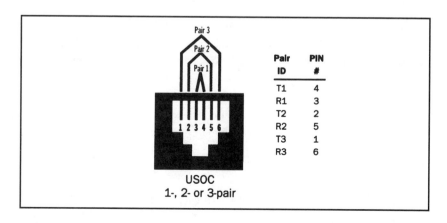

Pair ID	PIN #
T1	4
R1	3
T2	2
R2	5
T3	1
R3	6

USOC
1-, 2- or 3-pair

Figure 3-23.
RJ-14 jack.

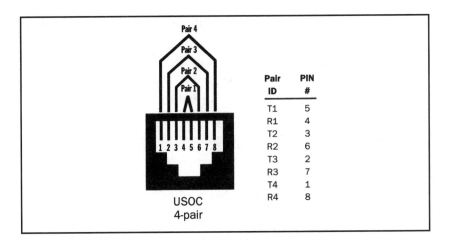

Pair ID	PIN #
T1	5
R1	4
T2	3
R2	6
T3	2
R3	7
T4	1
R4	8

USOC
4-pair

Figure 3-24.
RJ-45 connector.

this was an RJ-11 jack, only the first line would be shown. The second pair of pins may or may not be present depending on the manufacturer.

The plugs for this type of connector are normally crimp-on, using a special tool. The cable is usually flat non-twisted, or an adapter is used to accommodate round-twisted-pair cable. The jacks are connected with screw terminals or a type of insulation displacement connector.

The jack shown in *Figure 3-24* is an RJ-45 connector. It is shown with the normal USOC type of pinout allowing four telephone lines to be attached to this jack. Other special configurations are available to allow switching and special dialer hookup.

Still other pinouts are available and commonly used when this connector is used in local area network or other data applications. These configurations will be discussed in the Level II book under the section for data cabling.

Questions

1. Draw a diagram of a DPDT switch.

2. An SPDT switch has what form of contact arrangement?

3. What should be added to a relay to prevent failure of the driving circuit?

4. What is switched in an intercom using a voice call-in type of arrangement?

5. What is connected to pin 1 of an XLR connector, according to the AES standard?

6. What type of connector would most likely be used on an insert point of a mixer?

7. Name the three types of cable usually used with "F" connectors.

8. What is the impedance of a PL-259 connector?

9. What type of connector is usually used in video applications?

10. What is found on the outermost pins of an RJ-14 connector?

(Answers in Appendix A.)

SHOP ASSEMBLY

Introduction

Many systems contractors will have a number of different projects under construction at any particular time. These projects may be proceeding simultaneously, but be in various stages of development in response to job site installation schedules. These various projects may also be entirely different. One might involve a large-scale audio system for an auditorium, while another might be an intercommunications system for a school, or a nurse call system for a hospital. As we defined in the introduction to this course, a systems contractor's business involves many different integrated systems. The particular systems contractor with whom you are associated may have an entirely different profile than another contractor.

One of the interesting and challenging aspects of working in the systems contracting field is the diversity that you will encounter. Seldom do two systems contractors have identical profiles. Some may emphasize sound work, others may be more involved in security systems, while yet a third may be more involved in fire alarm and wired clock and time-control systems.

To the uninitiated, this can at first seem somewhat daunting and overwhelming: "You mean I have to learn all this?!" Not to worry. No one is going to expect you to be an expert overnight. The upside to this diversity is the sheer variety of work that you will be exposed to in this industry. Probably no other place where you could work will offer you the opportunity to learn and expand your knowledge in more areas than the systems contracting business.

There are a lot of very talented and knowledgeable people, both technicians and managers, employed in this industry. You will find the vast majority of them to be very helpful and willing to share their experience and knowledge. The old cliché is, "The stupidest question is the one that isn't asked." Ask questions — and if you don't quite grasp it the first time around, ask again. If you're new to this business, in all likelihood you will be working from detailed plans and under the supervision of an experienced technician. Many of the tasks will likely be repetitive, so once you have learned to do it the right way, just keep repeating the process.

Shop Assembly of Loudspeakers

As was said earlier, all systems contractors approach their business in different ways. However, in many instances, the newest members of the team will find themselves in the fabricating shop doing relatively simple repetitive assembly of devices. One of these tasks is the proper assembly of ceiling loudspeakers. We will study in the section under installing flush and surface-mounted speakers how a ceiling loudspeaker assembly is fabricated. Note that this type of fabrication is generally accomplished in the shop before the material is dispatched to the job site.

If these devices are purchased from the vendor in an assembled version, there is virtually no shop fabrication required. However, most astute contractors will inspect or audit these assemblies before sending them to the job site. This inspection may involve a 100% test in which an impedance bridge is used to determine that reading "sounds okay," whereas an audit generally denotes a random sampling of a certain percentage of the total. If the percentage test passes, then the shipment is assumed to be okay.

Conversely, if a set percentage of the audit fails the test, it is not uncommon to reject the entire shipment and return it to the vendor for correction. One of your tasks might well be to test incoming loudspeaker assemblies, to a prescribed standard.

If, on the other hand, the policy of your company is to purchase loudspeaker components separately from one or more vendors, then these will have to be shop-assembled. This type of assembly should be performed on a flat, preferably carpeted table. Make sure the work surface is clean, uncluttered and devoid of iron filings. Grilles can be very easily scratched, making them unsuitable for installation. Loudspeakers have a strong magnetic field and can attract iron filings that will prove very tedious, if not impossible, to clean. Likewise, some preliminary thought as to how and where you're going to store the assembled loudspeakers will avoid the possibility of further damage.

Note: The following procedure assumes that the components have been pretested or audited to ensure that incoming components are within specifications.

While there is no 100% correct sequence to assembling loudspeakers, most experienced technicians will use the following procedure:

1. Assemble the loudspeaker to the grille using the manufacturer's furnished hardware. Use caution that you do not torque (overtighten) the mounting screws, to avoid distorting the grille

or stripping the screw threads. Overtightening the screws holding the speaker can stress and deform the speaker basket, causing a misalignment of the voice coil. It also creates "dimples" in the face of the grille.

2. Most ceiling-type loudspeakers have a bracket for mounting the constant-voltage transformer. If a constant-voltage transformer is part of the assembly, mount the transformer to the bracket. Depending on the practice in your shop, this may involve mounting by using machine screws, lock washers and machine nuts or rivets. Use screws only long enough for secure mounting. Screws that are too long may penetrate the speaker cone during use.

    Most speaker transformer brackets will accommodate almost all transformers. If you run into the exception, advise your supervisor promptly. Don't get creative.

3. Solder the secondary winding (the voice-coil impedance winding) leads to the loudspeaker terminals. The secondary winding will usually be marked in ohms (Ω), and in most cases will be identified as 4 or 8 ohms. Exercise caution when soldering so you don't drip solder onto the loudspeaker cone.

    In some cases, the transformer might have terminals rather than leads. In this case, some jumpers of adequate length (plus a little slack) should be prepared in sufficient quantity for the job at hand. Good practice would be to make the negative lead one color and the positive another color. These jumpers will then be soldered to not only the loudspeaker but also to the appropriate impedance tap of the transformer.

4. Customarily, line matching transformer primary windings are marked in wattage taps over a typical range from .5 to 32W. These connections are made in the field and will not normally require shop attention. In some instances, all of the loudspeakers destined for a particular installation may have the same wattage

tap. In these cases, you might be asked to *dress* the unused leads. Dressing means that the unused leads should have exposed bare conductors cut off. Don't cut off any more insulation than necessary: someone might change their mind. Fold the end of the cut lead over on itself and tape or heat shrink the conductor. Preparation in this manner will avoid aggravating shorts once installed.

5. Test the completed assembly using prescribed procedures. This will generally involve connecting a signal generator to the primary winding and ensuring that the assembly operates properly. When you are out in the field doing installations, and have installed 125 loudspeakers in a ceiling only to find that some (or worse, ALL) of them don't work, you will recognize that someone at the shop didn't take the time to test for poor solder joints, shorted leads, open windings and frozen voice coils.

6. Store the assembled loudspeakers horizontally with a piece of cardboard or similar material between each stacked unit. Do not pack them vertically so the grille edge is the supporting member. A trip in a service van, however short, will virtually guarantee that a good percentage of speaker assemblies stored on their edge will not survive the trip.

Electronic Equipment Rack and Panel Assembly

Most commercial installations of sound, video and other electronics utilize equipment that is designed to be rack-mounted. Like many of the terms and equipment that we use today, the electronic equipment rack was borrowed from the telephone industry. Early on, the telephone companies found it necessary to have a convenient method of mounting their equipment. Thus, they developed what was known as a relay rack.

Relay racks were nothing more than two metal channels mounted on a base with provisions to attach the base to the floor. An example

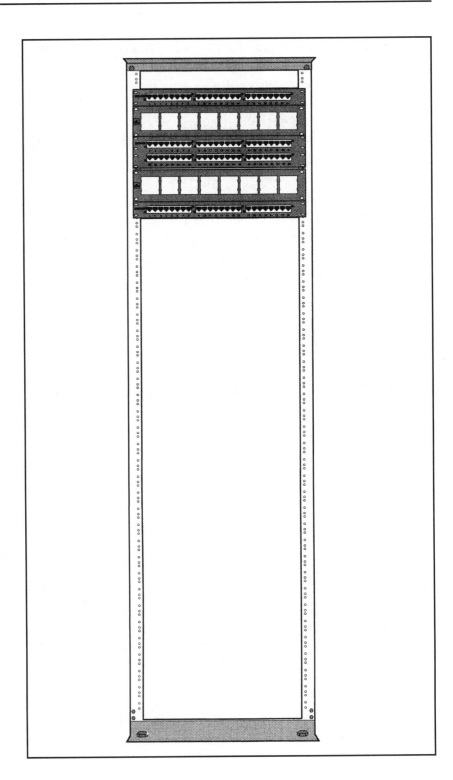

Figure 3-25. *Relay rack assembly.*

of a relay rack assembly is shown in *Figure 3-25*. Because much of the equipment common to that era was large and bulky, these rack assemblies were, by contemporary standards, massive.

Some of the early cinema sound system amplifiers (circa 1927), which were used to generate 10 watts of power, cost $10,000 in 1927 dollars and would occupy an equipment room the size of the room you're sitting in.

Inasmuch as these early rack assemblies were open, great care was taken to ensure that the interconnecting wiring was very carefully dressed, laced and formed to display a neat and workmanship-like installation.

The relay rack type of installation was to remain quite common up until the late 1950s and early 1960s. Small equipment boxes by companies like Bud began to be used for device mounting; but the enclosed, doored electronic equipment rack that we are accustomed to today simply didn't exist.

There are not many devices and components that are standardized in the electronics industry. Thankfully, the electronic equipment rack is one component that is standardized worldwide, thanks in part to the telephone industry.

A standard *rack unit* (RU) is standardized at 1.75" (4.45 cm). This is true whether you're working in North America or the wilderness of Siberia. Every subsequent RU is a multiple of the base unit. The following chart indicates the rack mounting standards in terms of RUs:

RACK MOUNTING (RU) MEASUREMENTS

Rack Units	Vertical Size (Inches)	Vertical Size (Centimeters)
1	1.75	4.45
2	3.5	8.89
3	5.25	13.34

4	7	17.80
5	8.75	22.23
6	10.5	26.70

Very few pieces of electronic equipment (with perhaps the exception of bulky audio and videotape or mixing equipment) measure much more than 6 RUs. In either case, the 1.75" increment is still the standard.

Also more or less standardized is the use of 10 x 32 machine screw drilling for rack assemblies. The 10 is a machine screw standard indicating the size of the screw, and the 32 indicates the number of threads per inch. This may vary depending on whether the rack assembly is going to be used in a metric country or in the United States.

Racks are standardized as to spacing in their vertical dimensions and in their horizontal dimension. The standard electronic equipment rack is sized to accept equipment with 19" (48.3 cm) horizontal dimensions. Some 24" (61 cm) and 30" wide racks are also available for special video, computer, telephone and surveillance equipment, sometimes with special mounting angles to hold 19" wide panels. This allows for cable distribution on the front face. There are also several "standard" rack depths, at 18," 22," 25," and 30."

Component manufacturers consequently produce equipment that will conform to and fit inside the EIA standard rack configurations. Again, exceptions to the rule include so-called consumer product VCRs, audio AM/FM receiver amplifiers, tape recorders, etc.

Manufacturers of consumer electronics still envision that their customers will pile this stuff in a stack or spend megabucks buying or fabricating some form of attractive housing.

Today, most equipment racks you will be called upon to use in fabricating a system will consist of enclosed cabinets with front and rear doors. In some cases, where multiple racks are used to form an overall system, the racks may be open-framed with no side enclosures. In this case, the multiple frames are bolted together (either in the shop or at the job site) and the side panels cover only the two exposed end frames (see *Figure 3-26*).

Fabricating the Equipment Rack

Normally, when called upon to mount equipment in an electronic equipment rack, the designer will have determined the proper placement of the equipment. The designer will have taken into consideration items like:

1. Heavy components (power amplifiers, for example), which will be mounted toward the bottom of the rack.

2. Equipment that requires user control, mounted for ease in adjustment.

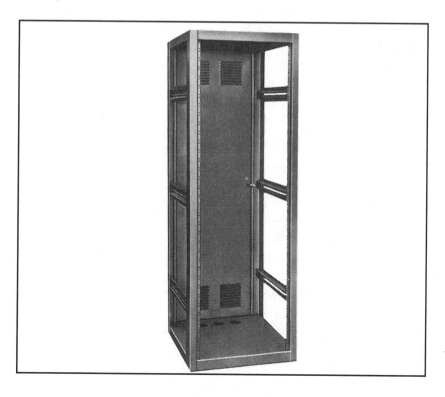

Figure 3-26. *Open-framed racks with no side enclosures.*

3. Ventilation panels and cooling fans positioned for maximum benefit.

4. Heat producing components given adequate spacing to eliminate component failures. The general rule is that the ambient temperature between component and chassis shall not exceed 50 degrees C.

5. Equipment laid out to minimize the length of the interconnecting wiring and maintain signal level separations. You don't run high-level speaker signal wiring in the same wiring bundle as low-level microphone signals. More on this later.

Consequently, in anything other than the very simplest of rack assemblies, a detailed drawing that shows where the equipment is to be mounted, and what ventilation panels are to be used and where, will be a part of the fabrication drawing schedule. These design considerations need to be followed meticulously.

Much of this type of drawing is done by using computerized design programs. An example of such a document is shown in the accompanying drawings.

How the assembly will be fabricated is primarily a factor of how it is to be transported to the job site for final assembly. In cases of multiple rack assemblies, it is hoped the designer took this into consideration.

One good way to install equipment in a rack (with a few exceptions) would be to have the rack laid down on a clean, well-padded floor and lay the equipment into the rack in this position. In this fashion, the weight of the equipment will hold itself into position without the necessity of holding the component in place (frequently impossible without two people) to attach it to the rack. This method will also allow you to make minor adjustments in spacing to accommodate equipment that might be just a tad out of tolerance.

To accommodate the various signal levels that will be routed through the rack, it is common to install an internal rack conduit system. This is frequently accomplished by using a plastic duct, which is often referred to by the brand name "Panduit." This material comes in various sizes and is adhered to the side frame of the rack. Again, the designer should indicate on the drawings where these various ducts are to be placed. Obviously, the ducts should be installed prior to mounting any of the components.

Similarly, there are often provisions for fans, blowers and supplementary AC wiring for service fixtures and lighting. All of these need to be installed prior to the placement of any of the rack's electronic components.

Once the components are laid into the rack assembly and properly positioned to line up with the rack mounting holes (an awl is a very useful tool in making this happen), then you can proceed to screw the components into the rack. Most component manufacturers provide mounting hardware with their components. Also, most suppliers of professional equipment provide cup washers to prevent the mounting screws from scratching the panel surface. If the vendor does not supply such washers, dip into your shop supplies and use them.

Equipment needs to be mounted securely; but not to automotive head-bolt torque. An electric screwdriver with a torque setting of 15 foot-pounds is quite adequate for most equipment mountings.

Wiring the Equipment Rack

Once all the equipment has been mounted into the rack, the wiring commences. Obviously, since most of the wiring will connect to the rear of the components, the rack assembly must be hoisted to a vertical position.

There are several different signal levels running about within the rack assembly. These are, in relative signal magnitude:

1. AC power current.
2. Loudspeaker distribution currents (+30 + 8 + 50 dBm).
3. Video transmission currents.
4. Intermediate-level signal processing currents (-20, +30 dBm).
5. Low-level, microphone or video input signals (-60 to -20 dBm).
6. Wireless or RF input signals.
7. Ground reference.

Any of these, if not adequately shielded or separated from dissimilar signals, can create cross-talk or induced voltages, and subsequent distortion to other signals.

A minimum of six inches is a good rule of thumb for separating wires within a rack. This rule cannot always be followed when shallow racks are used.

Again, the designer should have taken these factors into consideration and laid out the signal cabling diagrams with sufficient care. For the same reason, the installer should follow these diagrams to prevent problems from arising in the final system installation.

Proper grounding is also a particularly important subject in rack layout and wiring. There are literally volumes of materials that have been written on the subject of proper grounding. Suffice it to say that the designer should have determined what grounding scheme is going to be used within any particular enclosure or system. Again, the system design should be adhered to in order to prevent subsequent installation problems.

The procedures are applicable in both analog and digital circuits. However, it will be learned in subsequent lessons that the treatment of digital signals may require even more stringent measures to prevent cross-talk and subsequent distortions.

Equipment Terminations

Different types of electronic components come with different types of input and output connectors. Some have two or more types of connections to facilitate different user options. The most common connectors are:

Screw Terminals
These, as the name would imply, are simply screws which connect field wiring conductors to component circuits (see *Figure 3-27*). They can vary in size from 4 x 40" to 1/4 x 20", with the amount of current-carrying capacity corresponding with increased size.

When attaching wires to screw terminals, the conductor should be laid into the screw so that the conductor is rotated into the screw when it is tightened in its clockwise rotation. In the event the conductor is stranded, it may prove beneficial to tin (prepare the strands by soldering them together) the conductor strands into a solid conductor before attempting to connect it to a screw terminal. A poor mechanical connection to a screw terminal can create an intermittent connection that can be sheer misery to track down.

Some contractors and some specifications will require that conductors intended to be connected to screw terminals will be equipped with spade or lug connectors (See *Figure 3-28*). These are simply mechanical devices that are soldered or crimped to the conductor, which is in turn inserted under the screw terminal. There are specialized tools used to mechanically crimp the conductor onto the spade connector. When crimping such connections, it is

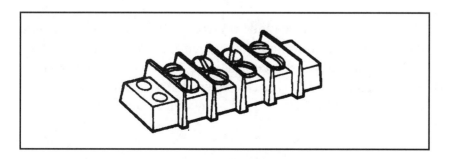

Figure 3-27.
Screw terminals.

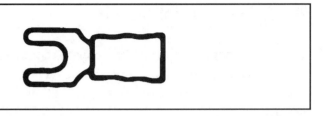

Figure 3-28. Spade or lug connector.

imperative that you use the manufacturers' specified instructions. Too many system failures have been attributed to connections that are improperly terminated. Don't ever use a pair of wire cutters to crimp this type of connector.

Component manufacturers will usually designate which terminals are positive (+) or negative (-). Unless the systems designer has indicated otherwise, positive terminals will attach to corresponding positive terminals, and negative likewise.

Screw terminal connections are generally used when there is little likelihood of the terminations being changed or revised on a frequent basis: "This power amplifier is always going to feed this speaker system array; this equalizer is always going to be connected to this power amplifier," etc.

Semipermanent Connectors
Semipermanent connectors fall into several different types. These type of connectors are used when it is anticipated that circuit configurations may vary on a regular basis.

By and large the most common connector in the audio field is the so-called XLR-type connector.

These connectors were (and are) used to facilitate quick and easy interconnection of components without the need for unscrewing and re-terminating screw connections. Most modern microphones, for example, use XLR-type connectors that permit them to be readily interconnected to various types of mixing equipment.

Phone plugs, or 1/4" phone plugs, are also used on some equipment to provide semipermanent connections. In most cases, the 1/4" phono plug is used to connect unbalanced circuits, and generally are employed in high-impedance circuit connections. The phone plug is a low-cost substitute for an XLR connector. It is generally used in lower-quality equipment, or where the manufacturer expects the equipment to be more often used in musical instrument (MI)-type applications.

The RCA phono jack is a very low-cost connector that consists of a central conductor. The shell or shield provides the ground connection. Principally found in low-cost consumer components and where unbalanced, high-impedance circuits are employed.

Preformed cables that use these types of connectors are available. However, in systems work, the preformed assemblies will be found to come in two varieties — too long and too short. In this case, you will probably have to solder the connectors to the connecting cable. This can be a tricky process, one that requires a little practice. Be prepared to destroy several RCA connectors before you get the hang of it.

As you get into more specialized elements of systems fabrication, you can expect to encounter additional types of connectors. Use only the correct connector and tool for each style of cable.

The BNC is used primarily in video and other high-frequency communications equipment.

The PL259 is a common connector in two-way radio communications and in some computer network applications.

The DIN multi-pin connector is found to be a common connector on computer-controlled equipment (see *Figure 3-29*). These can be found to vary drastically in pin count and connections, and the manufacturer's service manuals must be carefully examined to

Figure 3-29. *DIN multi-pin connectors.*

ensure that the pinouts are compatible between connected components.

This covers only a small sampling (but hopefully the major types) of connectors that you will encounter in fabricating and wiring systems.

Soldering Techniques

As mentioned several times in the previous discussions concerning connectors, good soldering is a must. The trick in good soldering is to get the material that is to receive the solder sufficiently hot to flow the solder. Never carry the solder from the iron to the solder surface. When the surface of the material is sufficiently hot, it will flow the solder onto the surface without any direct application of the iron to the solder.

There are also different types of solder. In most electronic connections, a rosin-core solder is the type to use. Conversely, a plumber will use a lead-acid solder for connecting copper pipes. The two are absolutely NOT interchangeable.

A good solder joint will have a bright, shiny appearance when cooled. A dull or gray appearance is an indication of a *cold solder joint*. A cold solder joint generally indicates that the conductor is floating under the solder flow and is not maintaining a good connection to the terminal to which it is connected.

One cold solder joint in an otherwise flawlessly fabricated system can absolutely destroy the integrity of the entire system. Remember when NASA had to shut down their entire space mission because of a faulty "O" ring? No system is better than its weakest link.

There are a lot of people out there who are more than willing to give you some degree of guidance. Ask the questions, and you'll probably be rewarded with helpful instructions.

Tools

Most systems contractors will have a collection of shop tools for your use. However, you will be expected (and will probably want) to provide your own hand tools.

Buy good quality tools and keep them properly maintained. A bargain-basement tool is bound to fail on you, and usually at the most inopportune and inconvenient time.

Experience will tell you which tools to add to your collection. For starters, the following minimum tools should get you started:

1. A good-quality leather tool pouch, designed to be worn about your waist, will save you a lot of steps and is the safest way to carry tools when on the job site.

2. Two or three different-sized blade screwdrivers of both the straight and Philips type. These obviously should have insulated handles and feel comfortable when you use them.

    A very handy screwdriver is the so-called *stubby*. These generally have a blade that is only 1-1/2" to 2" long, and are used when there just isn't enough space to fit a conventional-sized screwdriver. Stubbys also frequently have reversible blades, with a straight blade on one end and a Philips on the other.

3. A good-quality pair of needle-nosed pliers. These, too, should have molded, insulated handles.

4. A pair of side-cutter of diagonal pliers for cutting wire. Insulated handles are also a must.

5. As previously mentioned, an awl is a very handy tool.

6. A 10' or 12' tape measure. Most tool pouches have a pocket for a tape measure. Make sure the tool pouch you buy has such a provision.

7. A 6" adjustable (crescent) wrench.

8. A pair of adjustable gap pliers, sometimes called *slip-joint pliers*. These are very useful in making conduit nut connections and similar mechanical procedures.

9. A wire-stripper. These come in various types and sizes. Try these out and buy the one that you find the easiest to use. Some are sized for various gauges of wire, while others are of the type where the user sets the size by the amount of pressure applied to the device.

10. Finally, a good-quality pocket knife should round out your initial tool complement.

Keep your tools clean, sharp (when appropriate) and oiled. If a tool becomes damaged, replace it.

BASIC WIRING

Introduction

Most technicians working in the sound and communications industry are not licensed electricians. Their work as it relates to wire and cable falls in the low-voltage category. Few system contractors get involved in the installation of power distribution cable, and in many jurisdictions are precluded (by various licensing requirements) from installing AC distribution wiring.

As has been mentioned on several occasions, all systems contractors operate their businesses in different ways. In some locations and jurisdictions, the installation of cable falls within the scope of work for electricians. In other areas, the systems contractor will install all of the low-voltage wiring and cable that is associated with the scope of the project.

The procedures used for installing wire and cable are very similar, whether the cable will be used for power distribution or communication/signaling purposes. However, communications and signaling wire and cable have some special characteristics that necessitate handling and installation in a manner distinctly different from power distribution wiring.

In many cases, the type of wire and cable that will be used on a particular project is determined by the consultant, electrical engineer or designer. They will (hopefully) have taken into account the following factors:

1. The application of how and where the particular cable will be utilized.

2. Certain systems require specific wire and cable: i.e., TV distribution utilizes coaxial cable, etc. Microphone cable for balanced circuits must be two-conductor shielded. Obviously, direct burial cable must be very much different with respect to jacket composition than cable used in normal interior wiring systems.

3. What safety factors apply?

4. The provisions of the codes and standards governing the installation. The National Electrical Code (NEC), which is widely used by building codes, dictates in many ways what types of wire and cable are allowable for use under different circumstances.

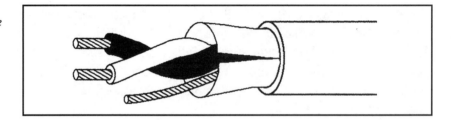

Figure 3-30. Multiple conductors enclosed in a jacket are referred to as cables, whereas a single conductor is a wire.

5. The electrical characteristics of the particular types of wire and cable that have been designed into the project.

Types of Wire and Cable

There are three general, broad categories of wire and cable that will be encountered in the sound and communication field. These are:

1. Metallic wire
2. Coaxial cable
3. Fiber Optic cable

The salient features of each type are discussed below. In this session, we will concern ourselves with metallic wire and coaxial cable and leave the detailed discussions involving fiber optic cable for a later section.

Metallic Wire
The most common type of wire found in fixed installation audio and signaling applications is metallic wire. This type of wire can be found in single-conductor and multiple-conductor configurations.

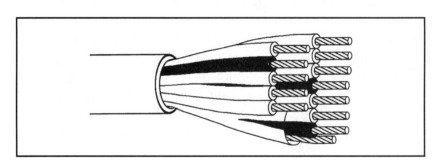

Figure 3-31. Metallic wire cable with stranded construction, composed of multiple small conductors woven into a single conductor.

Usually, multiple conductors enclosed in a common jacket (outer cover) are referred to as cables, whereas a single conductor is referred to as a wire (see *Figure 3-30*).

Wires and cables come with a variety of jacket types. A quick look into any manufacturer's wire and cable catalogue will show you a profusion of different types of jacket compositions, suitable for any number of specialized applications.

Metallic wire and cable can be of either stranded or solid construction. This refers to whether the metallic conductor is a single solid wire or is composed of multiple small conductors woven into a single conductor (see *Figure 3-31*).

Coaxial Cable
Coaxial is frequently associated with transmission lines involving television, video, radio frequency communications and data circuits (see *Figure 3-32*).

Different types of RF circuits employ various different impedances. Consequently, it is very important that the proper type of coaxial be specified and installed for any particular circuit design. The following chart shows the different characteristics of many of the more popular coaxial cables. *Figure 3-33* shows quad shielded coaxial cable.

Fiber Optic Cable
Appearing on the scene within the past ten years, fiber optic cable is a glass tube which carries transmitted signals as modulated light waves. Initially, commercial use of fiber optics was in telephone

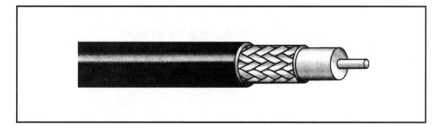

Figure 3-32. *Typical coaxial cable.*

189

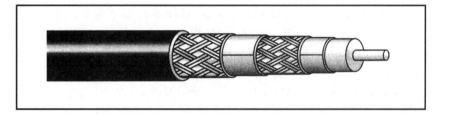

Figure 3-33. *Quad shielded coaxial cable.*

and data transmission networks. It has the advantages of wide bandpass, low noise and low line-loss characteristics (see *Figure 3-34*).

Fiber optic cable is being used in more and more applications as the cost of the product goes down and the installation connectivity procedures are simplified (see *Figure 3-35*).

Cable Installation Procedures

A little forethought before commencing a cable installation will save you a great deal of time and make the project a lot easier. There are five main considerations that will impact the installation of any cable pulling project:

1. Ambient temperature.
2. Equipment.
3. Conduit fill.
4. Mechanical fit in raceway.
5. Physical limitations.

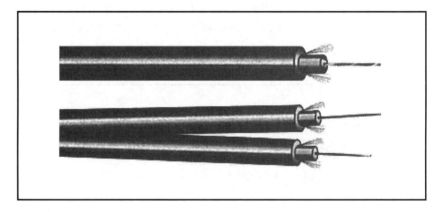

Figure 3-34. *Fiber optic cable.*

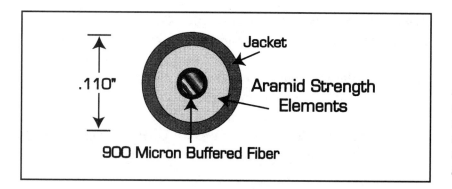

Figure 3-35. Fiber optic cable is used more as production costs go down and installation is simplified.

Mechanical Considerations

Before starting to pull wire or cable, check the following items:

1. Measure the actual length of the run of the conduit. Remember, the actual length of the installed conduit could be drastically different than what is shown on the plans.

2. Ensure that the lengths of cable you plan to install exceed the length of the run. Splices in conduit are frequently in violation of codes and specifications. Also, splices are costly and time-consuming to implement.

3. The conduit or cable tray should be deburred of all sharp edges to prevent damage to the cable jacket.

4. The conduit should be checked by pulling through a duct cleaning mandrel to ensure that construction debris is not present in the pipe. When necessary, the conduit should also be swabbed or alternatively "blown" to remove any accumulation of water.

Physical Installation of Wire and Cable

When installing cables, care must be exercised to avoid exceeding the *pull tension* for the types of cable being installed. For cables with copper conductors, the allowable pulling tension is equal to 40% of the breaking strength of the cable. Most wire and cable manufacturer specification sheets will indicate the break strength and maximum tension values for their various types of cables.

Exceeding the maximum tension value can cause internal cable damage by virtue of either stretching or breaking the conductors.

Spring scales and other tension-measuring devices can be used to ensure that the cable's tension level is not exceeded.

The terms "let-offs" and "cable payout" are used to describe the act of removing cable from spools or reels. When removing cable from reels, an arbor must be used so that the wire plays out from the revolving reel and not over the end (flange) of the spool. Allowing the cable to play off over the flange of the reel will twist the cable 360 degrees for each revolution around the spool. This will cause kinks. Kinks can damage the conductors and will definitely make it more difficult to install the cable. The problem with twists and kinks is multiplied when installing two or more cables from separate spools.

Cables are frequently pulled by using the conductors themselves attached to the pulling device (fish-tape, etc.). When pulled in this fashion, the maximum pulling tension is the sum of all conductors being pulled upon. Cables can also be pulled by using grips or clamps. One such grip is the "Kellems grip," manufactured by the Kellems division of Harvey Hubbell. These use the "Chinese finger puzzle" principal that distributes the pulling tension evenly across the cable assemblies.

On long runs, pulling compounds can be used to reduce the friction between the cable and the conduit surface. These can either be dry or liquid compounds. Make sure when using any type of pulling compound that it is compatible with the type of cable jacket being installed. Specialized pulling compounds are available when pulling fiber optic cables.

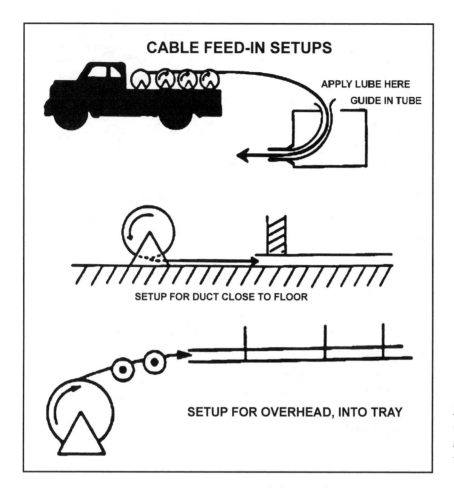

Figure 3-36.
Common setup
methods for installing
wire and cable.

Figure 3-36 (courtesy of Cablec Continental Cables Company, Marion, Indiana) shows some of the more common setup methods for installing wire and cable.

The feed-in setup should unreel the cable with a natural curvature (*Figure 3-37*) as opposed to a reverse "S" curvature.

Single sheaves may be used only for GUIDING cables. Arrange multiple blocks to hold bending radii whenever cable is deflected.

For pulling around bends, use conveyor sheave assemblies of the appropriate radius series (see *Figure 3-38*).

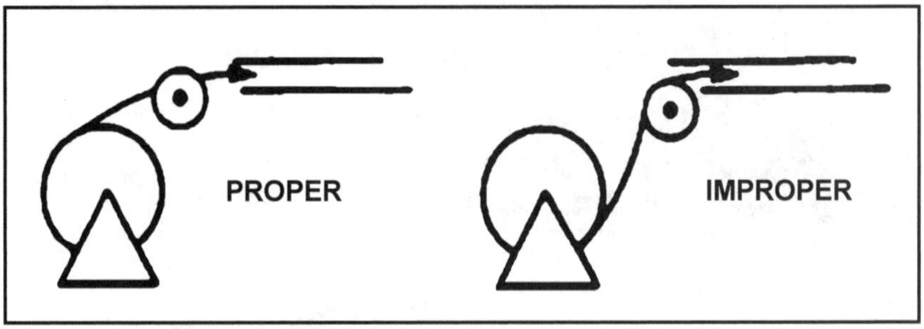

Figure 3-37. *Proper and improper feed-in setups.*

The pulleys must be positioned to ensure that the effective curvature is smooth and deflected evenly at each pulley. Never allow a polygon curvature to occur (*Figure 3-39*).

The fit of the pulley around the cable is also important when pulling heavy weights (e.g., pulleys at the top of a vertical drop).

Remember to use the radius of the surface over which the cable is bent, not the outside flange diameter of the pulley. A "10 inch" cable sheave typically has an inside (bending) radius of 3 inches! See *Figure 3-40* for more details.

Equipment

Make sure you have the necessary equipment and tools required to perform the cable pulling project. The equipment required for many wirepulling projects is listed:

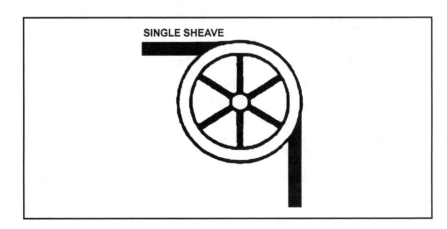

Figure 3-38. *Conveyer sheave assemblies are used for pulling around bends.*

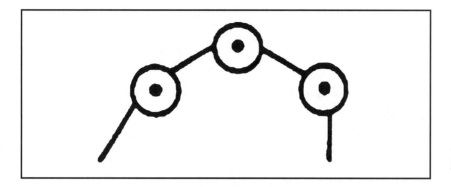

Figure 3-39. *Avoid polygon curvatures.*

1. Fish-tape or string blower/vacuum
2. Reel jack
3. Extension cords and GFCI
4. Radios or telephones
5. Hand lines
6. Rope slings
7. Hand winches (come-a-long)

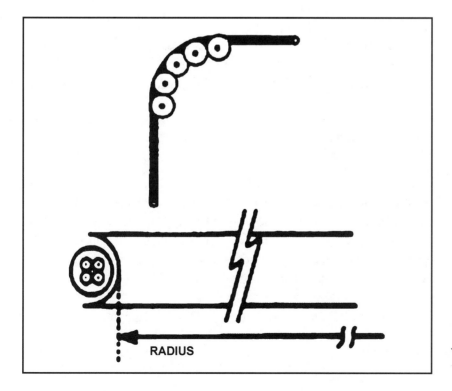

Figure 3-40. *Use the radius of the surface over which the cable is bent, not the outside flange diameter of the pulley.*

8. Swivels
9. Tension gauges
10. Lint-free rags
11. Prelubing devices
12. Cable end sealant (silicone caulking)
13. Metal files
14. Reel arbor
15. Reel brakes
16. Gloves
17. Flashlights or flood lamps
18. Duct cleaning mandrels
19. Shackles
20. Pull ropes
21. Basket grip pullers
22. Cable cutter
23. Cable pulling compounds
24. 50' measuring tape
25. Diameter tape
26. Normal hand tools

Depending on the particular project, not all of the items listed may be required. On the other hand, in other circumstances, additional items might be required to make the job go easier or have additional safety.

Conduit Fill

If the cable is going to be installed in conduit, the conduit must be sized in accordance with the National Electrical Code (NEC) requirements. Many of the wire and cable manufacturers list the conduit capacity for the various products which they manufacture. Cables which are not listed or for which information is not readily available in the field can be calculated as follows:

1. Measure the outside diameter (OD) of each cable.

    a. Square the OD measurement.

    b. Total the squares of all of the cables measured.

2. For pulls of three of more cables, multiply the total of step 1 by the factor 0.7854.

    a. The number thus arrived at is the total area of the cables in square inches.

    b. NEC requirements stipulate that a conduit having 3 or more conductors may not occupy more than 40% of the conduit (40% fill). A single conductor may occupy 53%, whereas 2 conductors are limited to 31%.

    c. Use the multiplier factor 0.7854 for three or more cables. Use the multiplier factor 1.0134 for two cables. Use the multiplier factor 0.5927 for one single cable.

3. Refer to *Table 3-2* and select the conduit size with a permissible area equal to or greater than the total area calculated in step 2.

4. The previous chart is based on the fill capacity as determined by the NEC.

    a. For conduit runs longer than 50 feet and not more than 100 feet, the installed number of cables should be reduced by 15%, or the next larger sized conduit should be used.

    b. Each bend in a conduit contributes to the pulling resistance. A derating factor is given in the following chart. This factor should be used as a multiplier of the straight run length of conduit to get a corrected conduit length:

Conduit Size	1/2	3/4	1	1-1/4	1-1/2	2	2-1/2	3	3-1/2	4	4-1/2
(Inches)											
Permissible Area	.12	.21	.34	.60	.82	1.34	1.92	2.95	3.96	5.09	6.38
(Sq. Inches)											

Table 3-2. *Conduit size chart.*

Total Degrees of All Bends	Multiply Conduit Length By:
45	1.2
90	1.3
135	1.4
180	1.5

c. If the conduit run exceeds 100 feet, or if there are more than two (2) 90-degree bends, a pull box must be installed.

Ambient Temperature

Frequently, wire and cable are installed in unheated buildings and in outdoor environments where the temperatures are not controllable. Low temperatures can be a concern when installing cable due to the increased brittleness of the jackets when exposed to low temperatures. The following chart indicates the minimum temperature when cable should be installed:

CP/EP-1C	-31°F
Welding	-41°F
N	-13°F
S	-76°F
CPE Jacket	-31°F
FREP, PE, XLPE-1C	-58°F
PVC	+14° F

CP...........	chlorosulfonated polyethylene (Hypalon®)
CPE........	chlorinated polyethylene
EP...........	ethylene propylene
FREP.....	flame retardant EP
N.............	neoprene
S.............	silicone
PVC.......	polyvinyl chloride

During cold-weather installation, cable should be pulled more slowly. At all times, care should be taken to ensure that the cable is not subjected to impact, drop, kinks or sharp bends. It is considerably more critical to observe these mechanical considerations when installing cable in cold weather.

Cable Connections

Obviously, cable is meant to connect something to something else. This could be a loudspeaker cluster connected to its associated power amplifiers, a nurse call master connected to its associated patient stations and staff/duty stations, or a video distribution network playing out from distribution amplifiers to numerous "taps" in a school classroom situation.

Many consultants and designers will stipulate that cable connecting field devices to "head-end" equipment cannot be directly connected to the head-end pieces of equipment. They will stipulate that field cable shall be terminated on connection blocks, and that jumper or umbilical cords shall tie the field wiring to the associated equipment. This is done for good reason:

1. The field wiring can be terminated and tested prior to the placement of the actual head-end equipment.

2. Testing and maintenance of the cable plant (the installed cable) can be more readily accomplished when the termination is on a logically identified terminal rather than on myriad pieces of equipment.

3. Future rearrangement of circuits is more easily accomplished when the field wiring is not carefully dressed and trimmed into equipment.

That is not to say that there are numerous cases where microphone cables are directly connected to their associated preamplifiers and loudspeakers are directly connected to their associated power amplifiers. However, as the project increases in complexity, the need for interfacing termination points becomes more obvious and more desirable.

This also brings up the subject of the proper labeling and documentation of circuits. During design, each cable pair or coaxial conductor run is designated to service a particular location or field device. It is therefore mandatory that the installed field cable(s) must be marked in some fashion to indicate which cable is meant to feed what device.

For example, let's consider a multizone paging system in which 6 amplifiers are going to feed 6 different zones for paging and background music. The cables connected to the loudspeakers associated with zones 1, 2, 3, 4, 5 and 6 are pulled back to the central head-end equipment rack. Unless these runs are properly identified, how will anyone (including yourself) ever know which runs are to be connected to which amplifier?

If the runs in this example are not legibly identified, how will any future serviceman (including yourself) ever hope to make sense of which cable runs are associated with the various zones? This simple example shows why proper labeling and documentation are a very important part of the cable pulling process.

Connections for cables and conductors can be made to screw terminals, punch down blocks (made popular by the telephone industry), or any number of solder and crimp connectors.

Cat 3 and Cat 5 wiring specifications introduce further methods of terminating cables.

We will cover more specific details on the manner for making mechanical connections of cable in later sessions.

Chapter 4
System Components

- **NICET Work Element 11010, Flush and Surface-Mounted Loudspeakers**
- **NICET's Description:** The applicant shall show an understanding of.... "Installation of speakers with proper support, back boxes, connections, and grilles. Utilize manufacturers' literature."
- **NICET Work Element 11011, Basic Microphones**
- **NICET's Description:** "Identify common microphone types, impedances and applications. Install microphones using proper mounting techniques."

Chapter Contents

FLUSH AND SURFACE-MOUNTED LOUDSPEAKERS

Introduction

For the purpose of this section we will confine our discussions to "typical" ceiling and wall-mount loudspeaker devices. Loudspeakers used in this type of application are generally cone-type loudspeakers, and come in 4," 6," 8," 12" and (sometimes but rarely) 15+" diameters.

Systems which comprise these types of loudspeakers are commonly installed to provide background music and paging service in commercial offices, retail, cafes/lounges/casinos, institutional, and to some degree, residential applications.

Depending on how the particular project was developed, the loudspeaker mounting locations may have been selected by the architect, electrical engineer, acoustical consultant, and/or the design builder. There are, as we shall study later, definite guidelines and considerations for the proper placement and density of various types of loudspeaker installations. However, you will probably find that the location where the loudspeakers are to be mounted is shown on the plans from which you are working.

Ceiling and Wall Building Materials

In general, there are two commonly used ceiling materials in most commercial spaces. These are "acoustic tile" and "sheet rock." Acoustic tile is almost always installed by acoustical tile contractors and is referred to as a "lay-in" or "grid" ceiling (*Figure 4-1*). A lay-in ceiling is built by hanging a "grid" of lightweight metallic channel which is supported by hanging support wires from the roof or structural members of the building. The grid is also supported by a special channel where the ceiling meets the walls of the room. Lay-in acoustical ceiling tiles usually come in 2' x 2' or 2' x 4' sec-

Figure 4-1. *Lay-in or grid ceiling.*

tions. These tiles are then laid in place and supported by the previously installed grid. The one definite feature of this type of ceiling is the ability to "push-up" the individual tiles to gain access to the space above the ceiling and below the actual roof of the floor above. This space between the suspended grid and the next floor is referred to by architects as the "plenum" space.

This space may or may not be a plenum as referred to in the NEC® code book. The NEC® code book defines a plenum as *"a compartment or chamber to which one or more air ducts are connected and that forms parts of the air distribution system"* (Section 100-A). Installation of speakers in a plenum requires special rated enclosures and wire, and conduit or wire fittings.

The acoustical tile ceiling is pretty much the preferred ceiling in today's commercial and retail spaces. Its popularity is due to the relative low cost of the material, the relative ease of installation, the ability to remove tile "panels" to access the plenum space, and the ability to modify the spaces for different uses in a fairly quick and low-cost way.

There are some variations on the acoustic tile ceiling. In most cases you will be working with a system as described above where the

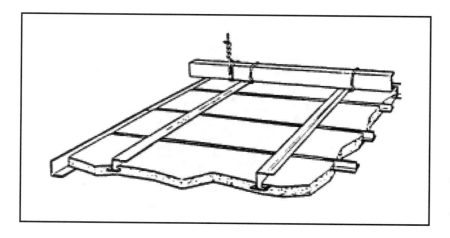

Figure 4-2.
Concealed spine
grid system.

grid is easily discernible from below the ceiling. However, there are some grid systems that are referred to as "spine" type, or concealed spine type systems (*Figure 4-2*).

When confronted with spine-type ceilings, proceed with caution. Without a knowledge of how the ceiling has been "keyed", a wrong move can bring down the whole ceiling structure in a very drastic and potentially dangerous and expensive fashion.

In some, particularly older, spaces you may encounter a "stick-on" acoustical tile ceiling. In this method of construction, the squares of the tile (usually 1' x 1') are glued directly to the ceiling structure. In such cases, you need to treat this type of construction in the same manner as you would a sheet rock ceiling.

Sheet rock, or gypsum board as it sometimes called, (Gyp board is another slang term for the material) comes in either 4' x 8' or sometimes 4' x 12' "sheets." The material also varies in thickness from a 1/4" to 3/4" with 1/2" or 5/8" being the more common size for commercial and residential buildings. Sheet rock can be used for both ceilings and walls. In both cases it is screwed or nailed directly to the wall studs or the ceiling joists. Once in place, sheet rock is impossible to remove without physically destroying the material.

Figure 4-3.
Loudspeaker with line matching transformer.

When this type of material is used on a ceiling and once the ceiling has been put in place it is said to be "closed-in," or sometimes the term "rocked-up" is used to describe this condition. If an access door or trap hasn't been provided, access to the space between the ceiling and the roof or overhead structure is virtually impossible.

Ceiling-Mounted Loudspeakers

A ceiling speaker assembly includes the following elements:

1. The loudspeaker, either by itself, or equipped with a line matching transformer (*Figure 4-3*).

Figure 4-4. *Backbox, or back can.*

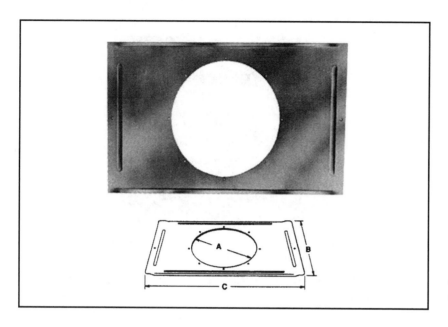

Figure 4-5. *A tile support.*

2. A backbox, or back can, of the appropriate size to hold the loudspeaker (*Figure 4-4*). These come in various depths and can be square or round. There are also backboxes that allow the associated grille to be held in place with torsion springs. Obviously you need to use the proper backbox to match the grille that will be installed later.

3. A tile support (*Figure 4-5*). This device is used to "span" or "bridge" the acoustical tile panel and keep the loudspeaker assembly from resting directly on the tile. When the tile is used as the primary support for the assembly, it has a bad tendency to sag.

4. A grille (or a baffle as it is sometimes, though technically incorrect, called) that is used to both connect the assembly together and to provide a decorative finish for the assembly (*Figure 4-6*). Grilles come in a variety of finishes, textures and colors. They may be fabricated from spun aluminum, painted metal, or plastic.

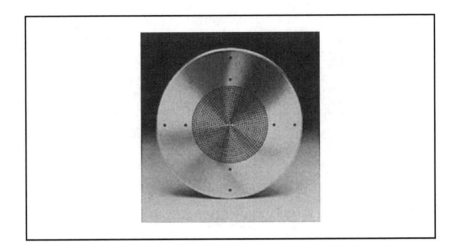

Figure 4-6. *A grille or baffle.*

Figure 4-7 shows how these parts are assembled together to form a final assembly.

In most new construction projects, the backbox is furnished to the electrical contractor and becomes part of his conduit (raceway) system. The electricians install these just like they install other electrical backboxes for other devices. In many cases they will also "hang" the backbox with support wires from the building's over-head structure. In most instances the acoustical tile installer or the sheet rocker will cut the holes to let the backbox protrude through the ceiling. This hole for the backbox should be the same (or a fraction larger) as the diameter of the backbox opening. In the case of a square backbox the same holds true; but, of course, the hole needs to be square.

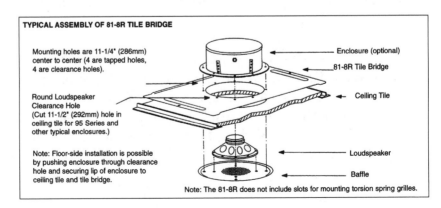

Figure 4-7. *Final ceiling-mounted loudspeaker assembly.*

As you can see from the illustration, the backbox has a "lip" or ring which is located 1/2" to 5/8" from the opening. This allows the backbox to "rest" on the tile bridge or on the sheet rock and come down through the tile or sheet rock flush with the finished ceiling.

Loudspeakers can be purchased separately with or without line matching transformers (if required) as well as with or without the grilles. Some contractors prefer to buy the grille, loudspeaker and transformer as a complete assembly. In this case the assembly, minus the backbox, is assembled by the manufacturer and shipped as a loudspeaker assembly.

If the option of buying separate components is taken, then the assemblies are generally put together in the shop. Transformers are mounted with self-tapping screws to the bracket provided on the loudspeaker frame, and then the loudspeaker is mounted on the grille. The mounting hardware for attaching the loudspeaker to the grille can be machine screws that pass through the grille and "pick up" the mounting holes in the loudspeaker and then are secured with lockwashers and nuts.

In this case, the screw heads will be exposed on the surface of the grille. Grille manufacturers generally furnish a hardware kit so that the exposed screw heads will match the finish of the grille. Other types of grilles have threaded studs welded to the interior of the grille which allow the mounting holes on the loudspeaker to set down flush on the grille. When the loudspeaker is seated to the grille, it is then held in place with lockwashers and machine nuts.

Whether the loudspeaker assembly has been purchased complete from a vendor or is shop-assembled, care must be taken in transporting the assemblies from the shop to the job site. Metal grilles are constructed of relatively lightweight metal and can very easily be bent and damaged in transit. Likewise both metal or plastic grilles can be easily scuffed in transit.

It varies from area to area, jurisdiction to jurisdiction and from project to project, but either the systems contractor or the electrician will install the proper field cable into the raceway. Again, in most cases, the wire is "looped through" the backbox with enough slack so that the wire can be cut at each loudspeaker location and subsequently attached to the loudspeaker, or if so equipped to the loudspeaker transformer.

Once the ceiling has been finished, i.e., painted, textured, etc., the loudspeaker installer connects the field wire to the loudspeaker (transformer). If a constant voltage line distribution is not being used, then the field wire is connected directly to the loudspeaker terminals. Care must be exercised so that they are wired in phase.

For example, assume a field cable having a red and a white conductor, where the red is designated positive and the white negative. The red (positive) should be connected to the same terminal on each and every loudspeaker. Some loudspeaker manufacturers mark their loudspeaker terminals with a "+" and "-", some use the convention "1" and "2". You would like the loudspeaker cone to move outward when a positive voltage is applied to the hot terminal. If you cannot identify the positive or hot terminal from manufacturer information, the speaker can be tested with a 9V battery to see which terminal, when connected to the positive pole of the battery, causes the loudspeaker cone to deflect outwards.

If this test cannot be made, at least connect all speakers in a similar manner.

By way of explanation, if you wire two loudspeakers "out of phase" then when one of the unit's cone is moving outward, towards the front of the loudspeaker, the other loudspeaker's cone is moving inward, or toward the back of the unit. The net result, if you stand between the two, will be a notable loss of sound and a poor frequency response.

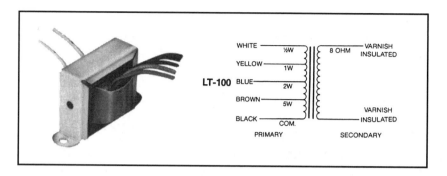

Figure 4-8. *Line matching transformer and schematic.*

Figure 4-8 shows a typical line matching transformer and its schematic.

Line matching transformers have, as can be seen, a primary and a secondary winding. The secondary is normally a voice coil impedance winding and connects to the loudspeaker. The primary has multiple "taps" usually marked in watts. The taps on a typical line matching transformer might be 0.5, 1, 2, 5 watts and common. Other combinations and values will also be encountered.

The primary winding tap of the specified wattage is then connected to the positive conductor of the field cable and the designated negative conductor is attached to the common terminal of the transformer. It is very common for transformer manufacturers to supply their transformers with wire lead connections. In such cases, the wattage tap associated with each lead is either stamped on the transformer frame, or the color-code of the leads is provided on a label attached to the transformer. In some cases, the transformer is not marked and you will need to refer to the manufacturer's data (specification) sheet for information.

We will discuss line matching transformers in detail in Level II.

In some cases, usually small one- to four-loudspeaker installations, it becomes necessary to connect the loudspeakers to the amplifier without using a constant voltage transformer. In these cases, the loudspeakers are wired in series or in a series-parallel configuration that will closely match the amplifier's output without exceeding its load requirement. In addition, wire should be up-sized to accommodate the extra attenuation experienced at these impedances.

Once the loudspeaker or loudspeaker transformer is wired it is then mounted into the backbox. The grille is secured with machine screws that pass through the grille and screw into the backbox. It is fairly common to use "tinnerman" clips as the securing device on the backbox.

Inasmuch as ceiling-mounted loudspeakers are mounted on a ceiling, the installation of these devices always involves working from a ladder or scaffold. The proper use and safety precautions should always be employed when using these types of tools. Refer to the section of this book under basic individual safety for a review of the proper use of ladders and scaffolds.

Surface-Mounted Loudspeakers

By definition, a surface-mount assembly is just as the name implies; a backbox mounted on the surface of a ceiling or wall. The procedure outlined above for installing flushed-in loudspeakers is basically the same for surface-mount assemblies.

Surface-mounted loudspeakers are used in spaces where there might not be a ceiling, such as a warehouse, a restaurant where the decor does not include a ceiling or in an existing space with a lath-and-plaster ceiling which the owner prefers not to have penetrated, etc.

An obvious consideration when installing surface backboxes is how to attach the backbox to the structure. The type of ceiling material is definitely important. Sheet rock ceilings lend themselves to the use of toggle bolts, for example, whereas concrete ceilings call for lead sets or a Hilti gun. Every installation of this type is going to require some thinking and planning before you make your first move.

Surface-mount devices can be very conspicuous — and if they're installed out of square, they become even more conspicuous. How many times have you noticed an exit sign that is improperly mounted? Sometimes it seems that installers of exit signs, and loudspeakers too, either don't own a ruler and level or don't know how to read them.

Wiring to surface-mount enclosures is also required; the use of surface raceways, or neatly stapled or fastened cable, will go a long way in presenting a finished and professionally installed appearance.

Wall-Mounted Loudspeakers

Most commercial loudspeaker assemblies that are used for ceiling-mount purposes will prove to be too deep to fit a standard 2" x 4" stud wall. In the past, the common practice was to use the "wood baffle" type of product. You've no doubt seen these devices in some older buildings. They were, and still are, a kind of slope-faced cabinet that hangs on the wall. Some of these are fabricated out of plastic rather than wood (see *Figure 4-9*).

Several companies market small, unobtrusive mini-cabinets that can be installed with a "U" bracket or on an Omni-mount™ bracket (*Figure 4-10*).

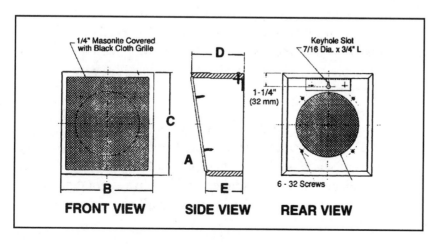

Figure 4-9. Slope-faced cabinet assembly.

Figure 4-10. *Mini-cabinet assembly.*

There are some flushed-in small speakers that will fit a standard 2 x 4 stud wall. These are used extensively in residential type applications due to their ability to blend in to that type of decor.

Any time you are installing an unfamiliar loudspeaker, read the manufacturer's suggested installation instructions. This can save you hours of time and sometimes the costly price of repairing a damaged wall or ceiling.

The more pressing problem in installing wall-mount devices (either surface or flush) is accessing the unit with the connecting wiring. Installations of this type rarely have a conduit system. Consequently, the wiring is generally run exposed and fed down through the walls. To access the space between the walls will require drilling the top plate on the stud wall and "fishing" the wire to the desired location.

Bear in mind that frequently a wood stud wall will have "fire blocks." Fire blocks are horizontal 2 x 4s that are installed between two adjacent studs, so before you poke a hole in the wall, make sure you can get the wire to that point. It is always good practice to scan a wall with a stud finder or to make all exploratory holes in the wall above the ceiling and only drilling the final location after you are sure the point can be reached.

Use of the cores in block walls is also a good way to conceal the wire. However, like the stud walls, all attempts should be made to verify accessibility before you drill your hole in the finished wall.

BASIC MICROPHONES

Introduction

The first practical microphone came into being as an integral component of Alexander Graham Bell's work on the development of the telephone. Also active in the field in a competitive fashion was Thomas Alva Edison's work on a practical telephone transmitter and receiver. Bell and Edison were to engage in some almost mutually destructive competitive practices, as they fought over the rights to provide telephone service A lot of this competitive action took place in England in the 1890s, and is not very well covered in U.S. history. Robert Clark, in his engaging biography of Bell, does an excellent job of reporting on the sometimes vicious legal and competitive practices that Bell and Edison used to try and get "a hand up" on the other in the telephone business.

Actually, Bell's first transmitter was an adaptation of J.P. Reis' loose-metal-contact transducer which he introduced in 1875. Reis' device was able to transmit tones of different frequencies but not intelligible speech.

The first microphone was Alexander Graham Bells' box telephone, into which Thomas A. Watson shouted and sang in the first intercity demonstration of the infant art of telephony.

As a side note, much of this early work was not an effort to develop a "telephone" but was directed toward developing a "harmonic telegraph". If you can imagine a world without a telephone, it did exist, and communications in the 1800s were confined to the use of telegraphy. The telegraph companies had a major problem in that only one message could be transmitted and received over one wire. Hence, to send multiple telegraph messages between, let's say New York and Chicago, they needed as many wires as they had messages which needed simultaneous transmission.

And, if you wanted to transmit from Chicago to New York at the same time - you needed another wire in the other direction. Obviously, the telegraph companies were anxious to develop a system which would allow simultaneous transmission in either direction over less wire. Bell was working on his version of the harmonic telegraph when he, by accident, stumbled over the principle of the telephone; and as his biographer notes was "a wise enough man to recognize his role as the father of his invention". The "harmonic telegraph" came into being much later and we recognize it as the open wire carrier and microwave transmission circuits of a latter day.

Also during 1875 (June 3rd to be exact) Bell achieved the ability to transmit intelligible speech with the use of a magnetic microphone. The idea worked, but not very well since it lacked sufficient sensitivity for telephone circuits. Berliner, who at the time worked for Edison, developed the loose-contact carbon transmitter. This proved much more practical for telephone usage.

After a few false starts, the *carbon button microphone* became the standard telephone transmitter. It was adopted circa 1897 and is still used as the transmitter component in the telephone to this day.

Engineers at Magnavox and Western Electric developed the concept of the *capacitance microphone* somewhere between 1914 and 1920. Credit for the invention of the capacitance (or electrostatic) transducer is generally given to Wente circa 1917. The instrument, however, was difficult and expensive to produce.

The need for a large and very expensive external power supply precluded the use of the device in any potentially commercial applications. Consequently, practical commercial capacitance microphones did not come into wide-scale general use until the 1950s.

The *electrostatic microphone* is frequently referred to (incorrectly) as a "condenser" microphone. Like the capacitor component, the electrostatic microphone, condenses nothing. Undoubtedly, it's a losing battle, but the modern "condenser" microphone is in reality an electrostatic or capacitance microphone.

After the advent of sound reinforcement systems and the birth of the radio industry in the 1920s, the race to develop better, more sensitive and more rugged microphones intensified. Cuttriss, Redding, Siemens, Wente, Thuras and Sawyer worked diligently on the design of ribbon, moving-coil, and crystal type microphones. Most of the modern principles of these various designs were achieved in the 1930s as the cinema and broadcast industries drove the demand for better and better type devices.

Microphone Types

As noted above there are several different types of microphones operating on varying principles. All of these designs date back to the time when most of us were not yet born.

Harry F. Olson of RCA Laboratories was to say in a paper presented to the AES in 1976, "The inception of radio broadcasting, electrical recording of disc records, and sound motion pictures, all in the 1920s stimulated the development and commercialization of high-quality studio microphones. The omnidirectional condenser and dynamic microphones were developed and commercialized in the 1920s. The bidirectional velocity microphone and the unidirectional microphones were developed and commercialized in the early 1930s. Microphones are still (1976) classified in these three basic types of directivity."

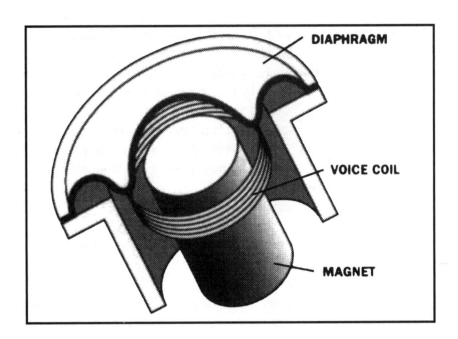

Figure 4-11.
Dynamic or moving
coil microphone.

There have been significant improvements in materials, fabrication techniques, and, of course, the introduction of transistor elements to replace earlier vacuum tube technology over the years. Better calibration methods and test instrumentation has allowed microphone manufacturers to produce more consistent yields in their products. But when you talk into a microphone that you bought yesterday, it is still the same basic device that Lou Burroughs in 1928 envisioned and started producing in his basement in Buchanan, Michigan — microphones for commercial production by his company — a company we now know as Electro-Voice.

There are essentially five different types of microphones. These, and their application characteristics, are as follows:

The *dynamic*, or moving coil microphone, operates on the principle that a coil of wire moved through a magnetic field will produce a varying electrical current corresponding to the movement of the coil. One of the most commonly used microphones; they vary in grade from low cost (paging) microphones up to very expensive devices used in recording and broadcast applications.

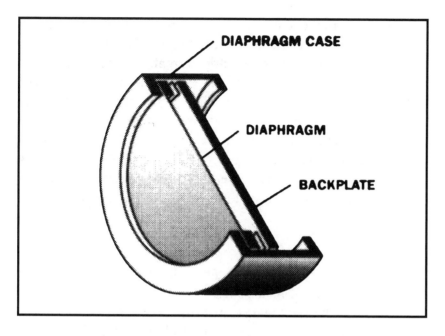

DIAPHRAGM CASE

DIAPHRAGM

BACKPLATE

Figure 4-12.
Electrostatic or
capacitance
microphone.

The dynamic microphone does not require a power supply for operation. A typical modern dynamic (moving coil) type microphone is shown in *Figure 4-11*.

An *electrostatic*, or capacitance microphone, operates on the principle that a light-mass membrane is one plate of a variable capacitor. The polarizing voltage is thus modulated by the varying capacitance. The electrostatic is generally considered to be the highest quality-type microphone available and widely used in recording applications. See *Figure 4-12*.

With the introduction of the electret type unit in the 1970s and the attendant reduced cost of manufacturing, the electret has found wide use in general sound reinforcement applications. The electret microphone has permanently charged elements and does not require a biasing voltage like early electrostatic microphones. However, because the output impedance of an electret microphone is very high and the signal level small, a voltage must still be applied to the microphone to power the impedance converter and preamplifier.

A *ribbon* microphone operates by placing the light mass, metallic ribbon in the center of a permanent magnet. As the ribbon is displaced, it causes a varying electric current to be induced into the ribbon. The ribbon microphone has very high-quality performance characteristics, but suffers from being very fragile. Mainly used in broadcast, recording and some cinema work, they are very seldom found in general-purpose sound reinforcement applications. By design, the ribbon microphone has a figure-8 or bidirectional pickup pattern favored by some performers and producers. Johnny Carson used a microphone of this type on his popular late-night TV show.

A *crystal*, or *piezoelectric*, microphone operates on the principle under which a crystalline structure will produce a small electric current when placed under stress. The stress is supplied by the attachment of the diaphragm to two similar crystals. A sound wave striking the diaphragm causes the crystal to be twisted, producing an electric current corresponding to the sound wave. A crystal microphone is a very high-impedance type of device. This characteristic severely limits the cable length that can be used between the microphone and its associated amplifier. It is extremely sensitive and produces an omnidirectional pattern. Generally, crystal

CHARACTERISTIC	OMNI-DIRECTIONAL	CARDIOID	SUPER-CARDIOID	HYPER-CARDIOID	BI-DIRECTIONAL
POLAR RESPONSE PATTERN					
COVERAGE ANGLE	360°	131°	115°	105°	90°
ANGLE OF MAXIMUM REJECTION (NULL ANGLE)	—	180°	126°	110°	90°
AMBIENT SOUND SENSITIVITY (RELATIVE TO OMNI)	100%	33%	27%	25%	33%
DISTANCE FACTOR (RELATIVE TO OMNI)	1	1.7	1.9	2	1.7

Table 4-1.
Microphone pickup patterns.

microphones are low-cost units associated with home recording and communications applications, and are very rarely encountered in sound reinforcement applications.

The *carbon* button type microphone operates on the principle of varying the resistance in a circuit to modify the current flow to correspond to the impressed audio signal. As noted earlier, the carbon microphone was the earliest type of practical commercial microphone. It was developed for use as the transmitter element in the telephone and is still used for this application. It must have a talk battery for operation and is not found in modern sound reinforcement applications.

Microphone Patterns

Microphones are said to have different types of "pickup patterns." This is a reference to their ability to accept sound from an audio source in a 360° angle around their position or to, in some measure, discriminate against sound developed to the rear or sides of the microphone (*Table 4-1*).

A microphone which accepts sound from a source anywhere in the 360° angle from which it is located is referred to as an *omnidirectional* type microphone.

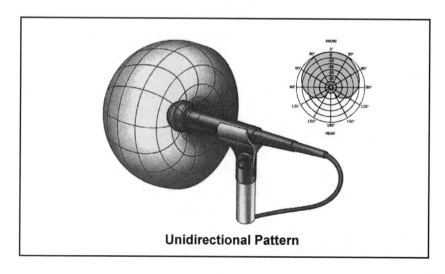

Unidirectional Pattern

Figure 4-13.
Cardioid or unidirectional microphone.

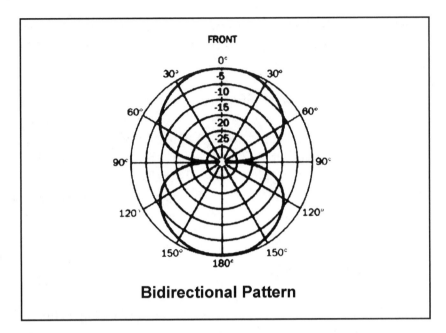

Bidirectional Pattern

Figure 4-14. Patterns produced by super-cardioid and hyper-cardioid microphones.

A microphone which discriminates against sound produced to the rear and sides of the unit is said to be a *cardioid* or *unidirectional* unit (*Figure 4-13*). Since in many sound reinforcement applications we may not want to detect sound from sources other than the talker or a particular musical instrument, we would use a unidirectional microphone. This type of microphone also provides some measure of protection against unwanted acoustical feedback.

The term *cardioid* comes from the "heart-shaped" pattern that this type of unit exhibits. Additional unidirectional patterns exist in the super-cardioid and hyper-cardioid microphone (*Figure 4-14*). These microphones exhibit even tighter patterns of control and have their null points, area of least pickup, more to the sides of the microphone than directly behind it as the cardioid does.

A special case is the pick-up pattern exhibited by the ribbon microphone. Since the unit's operation is dependent on the displacement of a metallic ribbon mounted in a magnetic field, it is susceptible to audio sources both in front of and behind the ribbon, but not to sources at 90 degrees to the ribbon. Hence, its pattern is in the

form of a "8." This was very convenient for panelists and inter-
viewers inasmuch as it allowed two people on either side of a desk
to use just one microphone, with equal pick-up from each source.

Microphone Impedances

Microphones are classified as either being "high-impedance" or
"low-impedance."

What does that mean?
A microphone can be thought of as an AC generator.

When you speak into a microphone you are creating an AC voltage
that corresponds in amplitude and frequency to the speech waves
(acoustical energy) that you are impinging on the microphone diaphragm. The diaphragm is a light-mass membrane that moves in
accordance with the acoustical signal which is presented to it. The
displacement of the membrane in turn reacts on a coil, the plate of
a capacitor, or a metallic ribbon to produce a very minute electrical
voltage.

You may recall from your studies of AC sources, that all generators
have some form of internal resistance. A microphone does not violate this principle. Consequently, all microphones have some degree of internal resistance which must be taken into account as
source output impedance of the microphone. For best operation, a
microphone should be connected to a load whose input impedance
is about 10 times the microphone's source impedance.

Low-impedance microphones will have a source impedance between 50Ω and 600Ω, with the most common impedance being
150Ω. These microphones should be terminated onto inputs with a
load impedance between 1kΩ to 2kΩ, for proper operation.

High-impedance microphones will have a source impedance of
25kΩ or greater, so the corresponding input would need a load impedance around 250kΩ.

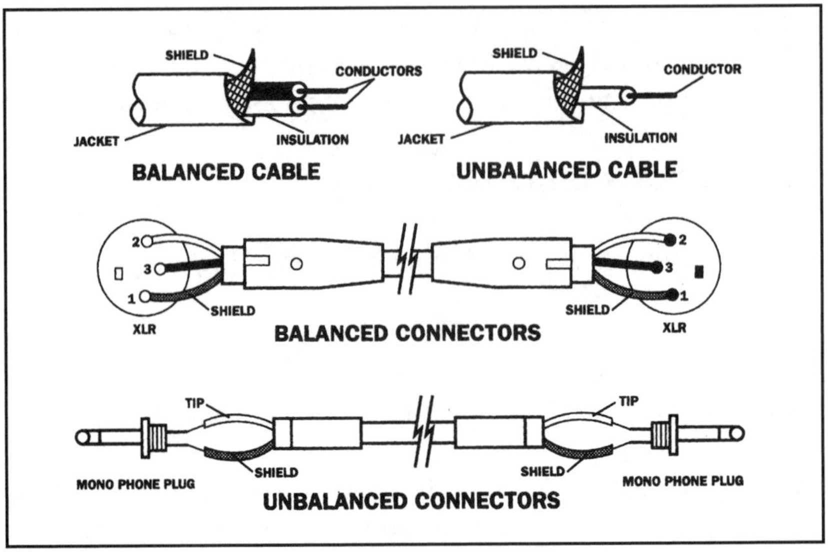

Figure 4-15. XLR *microphone connectors are often low-impedance while phono or RCA connectors are high-impedance.*

Interchanging microphones will result in poor operation in all configurations.

In general, low-impedance microphones can be physically recognized by the fact that their connecting cables (microphone cables) are two-conductor, shielded cables.

Because of the low impedance and the balancing of the microphone cable, low-impedance microphones can be placed a considerable distance from the mixer input without serious loss or degradation of signal.

In contrast, in general, high-impedance microphone cables use one central conductor with the shield being used as the other conductor. This is also a factor differentiating a balanced from an unbalanced circuit — but more on that later. High-impedance microphones are usually limited to cable lengths of 20 feet or less.

Almost always, a microphone connector of the XLR type is a balanced, low-impedance circuit; while a 1/4" phono plug or RCA type connector is an unbalanced, high-impedance circuit (*Figure 4-15*).

Screw terminals can certainly be a practical solution where the microphone connection to the device will seldom, if ever, be disconnected. Most preamplifiers will be logically marked as Hi, Lo, and ground (balanced or unbalanced) terminals.

Microphone Cable Wiring

Microphones are polarity-sensitive devices. Like a loudspeaker's cone, a microphone diaphragm moves in two directions in response to the audio energy present. If you have two microphones wired "out of polarity" and bring them into close proximity (like on a podium or lectern) the two units will cancel out a speaker talking into both of the microphones at the same time.

This is because one of the units' diaphragm is causing a negative-going signal as its diaphragm is moving inward (away from the grille) while the other microphone is causing a positive signal as its diaphragm is moving outward (towards the grille). Adding microphones of opposite polarity in a mixer will cause these signals to cancel each other out.

Since the XLR type connector was first introduced by Cannon, a debate has raged over which of the three pins should be considered positive (high) and which should be called negative (low). It really doesn't make any difference — as long as the convention is held consistent.

Finally, after 30 years of debate, the Audio Engineering Society (AES) issued a standard (S14-1992\.ANSI s.4.48.1991) that adopted the convention that pin 1 is ground (shield), pin 2 is hot (positive) and pin 3 is low (negative).

If standards are not to be followed, it is recommended that in any facility there be an agreement on which wire in a microphone cable is low or high and the adopted convention should be followed religiously. In a two conductor cable, the white, or lighter color, should be considered high and the black, or other darker color conductor should be considered low.

Microphone Physical Characteristics

Microphones come in a variety of sizes, shapes and mounting types. Some are designed to be hand-held, others are best used on a microphone stand, some are designed for use on podiums, while still others are designed to be used as lavalieres, which is probably why there are several hundred different types of microphones. The general classification of microphones includes:

Podium/Lectern Type
Used as the term would imply, on lecterns. Usually of the gooseneck type to allow them to be adjusted by various talkers (*Figure 4-16*). This type of microphone, widely employed in courtroom and legislative-type environments, can be either a capacitance or dynamic type with a unidirectional pattern. Normally, one microphone centered on the podium will be sufficient to pick up most talkers. However, if two microphones are placed on the lectern, they should be positioned together with their patterns crisscrossing to cover the most area with the least amount of interference.

Hand-held or Stand Mount Type
These are generally interchangeable and the selection rests more on the pickup pattern, the sensitivity, the physical size and the

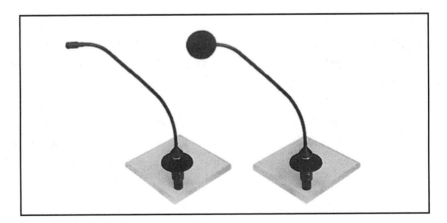

Figure 4-16.
Gooseneck
microphones.

Figure 4-17. Hand-held or stand mount microphones.

performer's or audio mixer's particular preference (*Figure 4-17*). Microphones, like loudspeakers, are a highly subjective entity. Some users are adamant about the sound of a particular microphone and insist that it is the only model they will consider. Working distance from the microphone and over system gain before feedback will be used to determine the proper pickup pattern of the microphone to be used. Unidirectional microphones exhibit much more gain before feedback than omnidirectionals, thus they are the microphones of choice for PA and sound reinforcement applications.

Figure 4-18. Lavaliere microphone.

Lavaliere Type

Lavaliere microphones are designed to be hung about the neck of the speaker, clipped to the tie or garment, or otherwise attached to the speaker's person (*Figure 4-18*). This provides the speaker using a lavaliere microphone much more freedom of movement while delivering his or her address.

Lavalieres are now mostly capacitance-type microphones with either an omni- or unidirectional pickup pattern. Again unidirectionals are the microphones of choice for PA applications.

Hardwired (cable-connected lavalieres) have practically disappeared, replaced by-and-large by wireless type devices.

Wireless Type

Wireless type microphones still use the same basic principles that we have studied up to this point. They will have a microphone capsule that might be a dynamic or a capacitance type element and have either an omni- or unidirectional pickup pattern.

Wireless microphones send their signal via a miniature radio transmitter (*Figure 4-19*). The audio characteristics of this type of microphone are no different than its wired equivalent, with the excep-

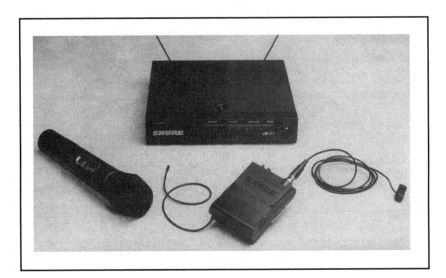

Figure 4-19. *Wireless microphones.*

tion that the microphone dynamic range is tied to the deviation allowed in the transmitter signal. Some wireless microphone systems use companding circuitry to restore full dynamic range. Essentially, a wireless microphone system substitutes a radio wave in place of a cable.

There are a number of other types of microphones which are employed for special applications. However, the above detailed microphones will be the ones that will be encountered in general use.

Phantom Power

Some capacitance-type microphones require some form of polarizing voltage. Many of the modern electret type use an internal DC battery to provide the necessary power for its impedance converter and preamplifier. However, a great many of these types of microphones are designed to receive their polarizing voltage from an external source. In the past this was usually an external power supply, but with the advent of low current-drain transistors it is quite common to see the "phantom power" supplied by the microphone preamplifier. This voltage is usually in the range of 9 to 48 volts DC and is supplied to the microphone over the microphone cable.

The term "phantom power" is a throwback to the early days of telephone and telegraph circuits. In order to derive additional circuits from existing open-wire telephone lines, the telephone companies and other private-line communication systems (railroads, pipelines, power companies, etc.) derived what was commonly referred to as a "phantom circuit."

The same principle is employed in delivering a phantom voltage to an electrostatic (capacitance) microphone. The voltage supplied by the mixer is divided equally across the microphone line. An identical current flows through both balanced conductors and consequently does not create any interference to the audio signal on the microphone "pair."

Most microphone mixers that provide phantom power are arranged with a switch to turn the phantom power on or off (sometimes on a channel by channel basis). Normally leaving the phantom power on even when it is being used to connect a dynamic microphone will not create any problems.

In the circuit in *Figure 4-20*, the phantom power is provided to the microphone line through two 6300Ω 1% resistors. Providing the DC power supply is of sufficiently low enough impedance, cross-talk between the inputs can be kept to usable levels.

The transformers shown in *Figure 4-20* are for providing a balanced input to the mixer and are not a part of the phantom circuit, although center-tapped transformers are sometimes used to provided phantom power in place of the resistors.

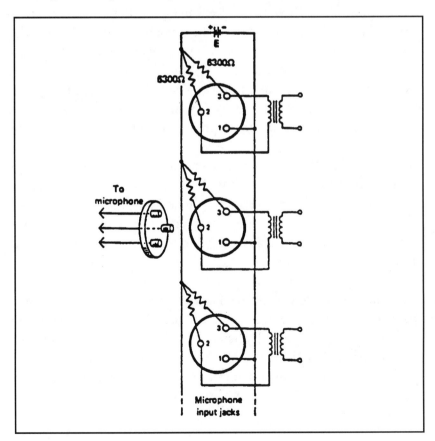

Figure 4-20.
Phantom power is provided to the microphone line through two 6300Ω 1% resistors.

Questions

1. Where can you find a definition for a *plenum*, as it applies to the system contractor?

2. What is commonly found to prevent the fishing of wires down a wood stud wall?

3. What is used to connect an 8Ω speaker to a 70.7V distribution system, allowing 1 watt of power to be applied to the speaker?

4. What should be used with a speaker when installing it into a typical drop ceiling?

5. Which way should a speaker cone move when a positive voltage is applied to the positive input terminal?

6. What type of microphone has a built-in figure-8 pick-up pattern?

7. What is the nominal coverage angle of a hypercardoid microphone?

8. What is the most common impedance, in ohms, of a low-impedance microphone?

9. According to the AES standard, which pin of an XLR-style connector is to be the positive or hot connection?

10. What is the DC potential difference between pins 2 and 3 on a microphone input with a 12-volt phantom power applied?

(Answers in Appendix A.)

Chapter 5
First Aid and Safety

- **NICET Work Element 11008, Basic Individual Safety**
- **NICET's Description:** "Follow standard safety practices in performing job tasks. Recognize and call attention to improper safely practices at the work site. (OSHA 2201 & 2202)."
- **NICET Work Element 11007, First Aid Procedures**
- **NICET's Description:** "Understand the basic rules and procedures of first aid. (See general handbook on first aid.)"
- **NICET Work Element 11009, Personal Protective Equipment**
- **NICET's Description:** "Properly use safety equipment such as eye protection, gloves, hard hats, hearing protection, safety shoes, etc."

Chapter Contents

BASIC INDIVIDUAL SAFETY
Introduction
Strains
Watch Where You're Going
Dress For Safety
Use the Proper Tool for the Job
Ladders and Scaffolding
Power Tools
Powder-Actuated Tools
OSHA Standards
Additional Regulations

FIRST AID PROCEDURES
Introduction
Shock
Heart Attack
Simple Fainting
Fractures and Sprains
Burns
Heat Exhaustion
Heat Stroke
Wounds
Stoppage of Breathing
Foreign Body
Poisoning By Mouth
First Aid Kits
Company Safety Officer

PERSONAL PROTECTIVE EQUIPMENT
Introduction
Personal Safety
Eye Protection
Hearing Protection
Foot Protection
Hand Protection
Questions

BASIC INDIVIDUAL SAFETY

Introduction

A systems contractor's employee frequently will find himself or herself working on job sites. These projects almost always involve workers from other trades, all of whom must work together on their various tasks simultaneously. To insure safety, all of the workers must cooperate and coordinate their work. No one trade has precedent over another and the safety of all workers should be the major consideration of all involved. The goal is to get the job done in a coordinated and safe fashion.

Millions of work days per year are lost because of on-the-job injuries. Not only are these expensive, they cause untold pain, suffering, and inconvenience not only to the individual worker, but to the families and the associates of the injured party. On-the-job accidents need not occur if individuals will take a little extra measure of care, pay attention to their working conditions, and avoid taking needless risks. Accidents occur because of people's carelessness or violation of basic safety rules.

Much has been done to improve job safety. Protective equipment is greatly improved and more comfortable and convenient to use, tools have been made more protective for the user and the sheer appreciation for the need for worker safety has helped immeasurably to improve the on the job safety record.

Ultimately, it is up to you as an individual to practice safe work habits. If you follow the safety rules, you will make the workplace safer for both yourself and your fellow co-workers.

Strains

Learn how to lift materials. If an object is too heavy to lift by yourself, get some help. Lift with your legs, not with your back.

Watch Where You're Going

Keep your workplace clear of unnecessary materials and tools.

Don't let materials and tools accumulate underfoot and create a dangerous situation where you could trip or stumble.

Stepping backward without checking your position is a sheer invitation to disaster. A false step on a ladder or scaffolding can result in a severe or fatal accident.

Dress For Safety

Be sure you have an approved safety helmet and goggles when you report to the job site.

Wear rubber-soled safety shoes to help you keep your footing. Open-toed shoes or sandals are meant for the beach — they have absolutely no place on a job site. If you show up on a job site with this type of footwear, expect to be sent home without pay.

Don't wear baggy or loose fitting garments. These types of garments can catch on things and tools and cause unnecessary accidents. The same is true of rings and loose jewelry.

Wear a tool pouch or kit. Don't carry tools in your hip pocket or try and climb a ladder with a handful of tools.

Use the Proper Tool for the Job

Many cuts and lacerations are caused by trying to make do with the improper tool for the job. Make sure your tools are maintained in proper fashion; chisels should be sharp, hammers should not be mushroomed or have splintered handles. Screwdrivers should not be rounded over or be too blunt for the project at hand.

Ladders and Scaffolding

Much of the work performed by systems contractors' employees will require the use of ladders and scaffolding.

The potential for serious accidents is greatly increased when ladders and scaffolding are involved. It is imperative that you know and follow the safety rules for using ladders and scaffolds.

1. Any ladder with missing rungs or steps, broken or split side rails, or other construction defects should be removed from the job site immediately.

    If you don't trust it — don't use it!

2. Always set a ladder so that its base is one foot from the wall for every 4 foot of vertical height.

3. Tops of ladders should extend a minimum of 36" above the upper landing on which they are resting.

4. Ladders should have their base substantially located on a firm solid surface.

5. Secure the upper rails of a ladder so that it will not displace and cause a serious fall.

6. Do not use metal ladders for electrical work or where the ladder might come in contact with electrical conductors.

7. Don't set ladders or scaffolding in doorways or passageways unless you provide some form of barricade or other suitable guard.

8. Hold onto both side rails when ascending or descending a ladder. Don't carry materials or tools in your hands when climbing a ladder.

9. When using scaffolding, insure that the footing for the scaffolding is level.

10. *Never* jack up scaffolding with boxes, bricks or concrete blocks.

11. Lock *all* casters to prevent movement and use outriggers when suggested by the scaffold manufacturer.

12. Employ toe boards and guard rails on scaffolding to prevent an accidental fall.

13. Properly brace all scaffolding in accordance with the manufacturer's directions.

    If it doesn't feel safe — don't climb it!

14. The height of a scaffold must not exceed four times the minimum base dimension. If the scaffolding you are using has a platform of 4' x 8'; the working height should not exceed (4 x 4) or 16.'

15. Move scaffolding only when the floor is level and free of obstructions. Push the scaffold by holding it as near to the base as possible. Do not allow anyone to "ride" the scaffold during movement.

Power Tools

1. Electrically powered tools should be of the approved dual insulated and/or properly grounded type. Third wire "ground" connections should never be compromised. OSHA requires that all power tools used on a job site should be plugged into a circuit provided with ground fault interrupter protection. If the circuit is not provided with GFI circuit breakers, portable in-line protectors must be provided by the contractor.

2. When using power tools, follow the manufacturer's instructions as to how to operate them properly and safely.

3. Guards on power tools are there for your safety. Don't incur unnecessary risk by defeating these guards.

4. Do not operate electrical power tools while standing on wet surfaces.

5. Do not operate power machinery with exposed gears, straps or other moving parts unless such moving parts are adequately protected.

6. Pay particular attention to electrical service cords and extension cords.

   a. Electrical cords should not have unapproved splice.
   b. Cords should not be frayed or worn.
   c. Extension cables should be protected from traffic and accidental damage.
   d. Don't hoist or lower power tools by their electrical cords

7. Defective power tools should not be used until they have been repaired.

Powder-Actuated Tools

Tools of this nature are actuated by a shell or charge. They are used extensively for shooting connectors and terminal hangers into concrete. Tools of this nature improperly used have a huge potential for causing injury to not only the user, but to other workers in the immediate area:

1. Only persons trained in the operation of powder-actuated tools shall be allowed to operate them.

2. Protective eye wear (goggles) must be worn at all times when using those types of tools.

3. Loaded tools shall never be left unattended.

4. The tool shall be cleaned oiled and tested everyday. Special attention needs to be taken to insure that the safety checks on these tools are functioning properly.

5. Tools of this nature need to be stored in their cases and be placed under lock and key.

6. Using this type of tool to drive a pin or fastener into easily penetrated materials must be avoided. It is permissible to use a powder-actuated tool in this manner only if the penetrable material is backed up with a substance that will prevent the pin or fastener from becoming a flying missile hazard on the other side of the material.

OSHA Standards

Many of the provisions of the Occupational Safety and Health Act apply to construction sites. You may be working on a job site that is visited by an OSHA inspector. This inspector's job is to check the work area for any unsafe working conditions that might violate OSHA standards. When this happens, extend the inspector every courtesy and immediately notify your foreman so that he can represent your employer in the inspection.

One of the first things the inspector checks is the proper posting of OSHA form 2203, Right-to-Know Law information, Material Safety Data Sheets and a safety plan.

OSHA's rules and standards are for your benefit. Know and obey its rules at all times.

Additional Regulations

In addition to the federal OSHA standards, rules and regulations, local cities, counties and states may very well have laws and ordinances regulating work place safety. OSHA is a minimum standard. States and municipalities can, and in many cases have adopted regulations that are more stringent than the OSHA standards.

You may also find that the client on whose property you are working may have special rules and procedures that you must follow. A good example of these client imposed rules are the entry permission procedures to refineries and chemical processing plants. You're on their property and they make the rules.

FIRST AID PROCEDURES

Introduction

The following basic information is drawn on information contained in a handbook on first aid provided by the Greater Milwaukee Chapter of the American Red Cross.

Shock

The symptoms of a person in a state of shock are: Pale, clammy skin, sweating, weakness, dizziness, nausea, abnormal pulse and breathing rate, dilated pupils, thirst, confused behavior.

If someone displays the above symptoms, take the following action:

1. Lay the victim down. If no spinal or leg fractures are suspected, elevate feet and legs 8 - 10". If victim is having trouble breathing or has head injuries, and you do not suspect neck or spinal injuries, elevate head and shoulders instead of the legs.

2. Maintain body temperature, but do not overheat the victim. Keep the victim just warm enough to prevent chilling. Too much heat at the surface will draw blood away from the vital organs.

3. If the victim vomits or is bleeding from the mouth, place him/her on one side to allow the fluids to drain and to maintain an open airway.

4. Call 911 (or the local emergency rescue number).

Heart Attack

The symptoms of a person suffering a heart attack are discomfort or pain in the center of the chest behind the breastbone. A victim may describe the pain as uncomfortable pressure, squeezing, tightness, aching, crushing, constricting, oppressive or heavy. The pain may also be present in one or both shoulders, arms, neck, jaw or back. Other symptoms include shock, sweating, nausea, weakness, shortness of breath.

If you observe someone displaying the above symptoms it is *essential* that you seek medical assistance *immediately*. Time is of the essence in cases of a possible heart attack:

1. Call 911 (or the local emergency rescue number).

2. Have the victim stop whatever he/she is doing and have them sit or lie down in a comfortable position that allows them to breathe most easily. *Do not let the victim move around.* Activity makes the heart work harder.

3. Monitor airway, breathing and circulation. If the pulse stops, give CPR if you know how.

4. Loosen restrictive clothing.

5. Get information from the victim or bystanders (name, age, previous medical problems, where it hurts, how long the attack has been, and what kind of pain it is).

Simple Fainting

1. If a person feels faint, suggest that they lower their head between their legs.

2. If the person actually faints, lie him/her down with the feet elevated until they recover completely.

3. Give first aid for shock.

Fractures and Sprains

A victim of a fracture may have heard a pop, snap, grating, or tearing sound in the area affected. There may be swelling, discoloration, deformity, local tenderness, loss of function in the affected limb.

Take the following actions:

1. Call 911 (or local emergency rescue number).

2. Do not allow victim to move the injured area.

3. If the fracture is open, use a clean dressing to cover the injury and control the bleeding. *Do not attempt to replace the bone(s).*

4. Treat for shock.

5. Splint the injury if professional help is not readily available.

6. In general, place the injured limb in a normal or comfortable position for the victim, unless:

a. The maneuver causes significant increase in pain.

b. You must apply force to move the area.

c. There is any resistance to straightening the injured limb.

d. The fracture might involve a joint.

7. Check the pulse and circulation below the fracture site before and after splinting. If there is no pulse after splinting, you may have applied the splint too tightly. Loosen the splint and check the pulse again.

Burns

A burn victim may suffer 1st, 2nd, or 3rd degree burns. The symptoms of each of these are:

1st degree: Redness, mild swelling, pain.

2nd degree: Deeper tissue damage, blisters, either closed or open ("wet" appearance), redness, swelling, pain.

3rd degree: Deep tissue destruction, white or charred appearance, complete loss of all skin layers, pain may or may not be present.

Take the following action(s) appropriate to the degree of the injury:

1st or 2nd degree burns with closed blisters

1. Flush with cool water until the pain subsides.

2. Apply loose, moist, sterile dressing, and bandage.

3. Care for shock.

2nd or 3rd degree burns with open blisters.

1. Apply loose, dry, sterile dressing and bandage.

2. Care for shock.

3. Check and monitor airway breathing and circulation.

Chemical Burns

1. Flush immediately with large amounts of water for 15 to 30 minutes.

2. Have someone call 911 (or local emergency number).

3. Remove any affected clothing or jewelry.

4. Cover with loose, dry, sterile dressing and bandage.

5. Care for shock.

6. Check and monitor airway, breathing and circulation.

Electrical Burns

1. Check to make sure the power source is disconnected.

2. Check airway, breathing, and circulation.

3. Call 911 (or local emergency number).

4. Check for multiple burn sites.

5. Cover with loose, dry, sterile dressing and bandage.

6. Care for shock and monitor vital signs.

Heat Exhaustion

A person who is suffering from heat exhaustion will display the following symptoms: Profuse sweating, cool, pale moist skin, dilated pupils, dizziness, weakness, nausea, possible vomiting, headache.

1. Remove victim from heat and move them into a cooler environment.

2. Place victim on his/her back, feet up (shock position).

3. Remove or loosen clothing. Cool victim by fanning vigorously and by applying cool wet cloths or cold packs (place a dry cloth between the cord pack and the victim's skin).

4. Give care for shock.

5. If the victim is fully conscious, give them one-half glass of water every 15 minutes, as tolerated.

Heat Stroke

A person who has suffered a heat stroke will display the following symptoms: hot, red skin, wet or dry skin depending on the person's activity, high body temperature, constricted pupils. The person may be weak and uncoordinated, may experience a headache, may be nauseous, may lose consciousness and may experience seizures.

1. Call 911 (or local emergency number).

2. Get the victim out of the heat and into a cooler environment.

3. Cool the victim fast. Keep the body wet (use cool wet cloths, DO NOT use ice). Fan vigorously.

4. Give care for shock.

5. Give nothing by mouth.

Note: If the victim begins to shiver, stop treatment and cover victim until shivering stops. Remember shivering produces more internal body heat. When shivering stops, resume treatment, but use warmer cloths and water.

Wounds

1. Control bleeding. If severe bleeding is present, apply direct pressure on the wound. If bleeding cannot be stopped, apply pressure to the supplying blood vessel, i.e., pressure point. Get medical help.

2. Cleanse minor injuries thoroughly, using plain soap and water after washing your own hands.

3. Apply dry sterile dressing and bandage snugly into place.

4. If infection is evident, get the victim to a physician promptly.

5. Treat for shock.

Stoppage of Breathing

If a person stops breathing, perform the following steps for mouth-to-mouth respiration:

1. Tip the victim's head back and look, listen, and feel for an exchange of air.

2. If the victim is not breathing, pinch the nostrils, make a tight seal over the victim's mouth with your own and give a full-size breath.

3. Breathe once every 5 seconds for an adult.

4. In the case of infants or small children, cover both their mouth and nose with your mouth and give small puffs of air from your cheeks, breathing every 3 seconds.

5. Call police, fire department, rescue squad, hospital or a physician.

6. Treat for shock.

Foreign Body

If a person gets a foreign body in his or her eye:

1. Do not rub or let the victim rub the eye.

2. Try to make the foreign body lodge on the inside of eyelid.

3. Roll back eyelid and remove object with the corner of a handkerchief or tissue.

4. If the foreign object is embedded, close the eye and bandage shut and seek medical attention.

If a person gets a foreign body in an air passage — which might occur while eating and the victim starts to choke:

1. Activate EMS immediately by sending someone to call 911 (or the local emergency number) if the victim is still conscious and unable to cough or speak.

2. Stand behind the victim, wrap your arms around him/her, then place the thumbside of one of your fists into the abdomen above the navel but below the ribs.

3. Give four inward and upward thrusts.

4. If ineffective, repeat the sequence.

5. If ineffective, open the victim's mouth and using the index finger of one hand, sweep across the back of the victim's throat from one side to the other and remove any foreign matter.

6. If these maneuvers are ineffective, repeat the procedures.

Poisoning

The symptoms that might indicate that a person has suffered poisoning by ingestion are: opened containers or plants nearby, nausea, vomiting, diarrhea, abnormal breathing and pulse, unusual breath or body odor. There may be burns around the mouth. The victim may also display drowsiness, unconsciousness, or convulsions.

1. Take any containers or plant material to the phone.

2. Call 911 (or local emergency number).

3. Call poison control center and follow their instructions.

4. Care for shock.

5. Monitor vital signs — airway, breathing and circulation.

6. *Do not* give anything by mouth unless the poison control center or a medical professional tells you to do so.

7. Save any containers, vomitus or materials for EMS. These can help to identify the poison.

In a case where a victim of poisoning has inhaled some poisonous substance the symptoms might include: Dizziness or headache, unconsciousness, difficulty in breathing, discoloration of the lips and mucous membranes.

1. Survey the scene. *Do not* enter the area unless it is safe to do so.

2. If safe, remove the victim from the area.

3. Monitor the victim's vital signs, breathing, airway, and circulation.

4. Call 911 (or local emergency number).

In a case where a victim has been poisoned by absorbing some poisonous substance the symptoms might include: skin reactions, itching, eye irritation, abnormal breathing and pulse, headache.

1. Remove the victim from the source of the poison

2. Remove affected clothing.

3. Flush all affected areas with lots of water.

4. Call 911 (or local emergency number).

5. Call the poison control center and follow their directions.

6. Care for shock.

7. Monitor the victim's vital signs.

First Aid Kits

Every shop and all of a company's service vehicles should be equipped with a properly stocked and maintained first aid kit. These should be inventoried on a routine basis and restocked or refreshed as required.

Company Safety Officer

Each organization needs to appoint one responsible individual to serve as the company's safety officer. The responsibility for the conducting of safety training and maintaining of safety records lays with this appointed individual. Not only is it good common sense to make this move, but in the event of a serious accident to an employee, it will indicate to the investigating authorities that your company has an established procedure for handling safety issues.

PERSONAL PROTECTIVE EQUIPMENT

Introduction

OSHA has established regulations governing the use of personal protective equipment. These are covered under section 29 CFR 1910.132 — .138. Because your health and welfare are important to both you and your employer, take heed of the following points:

1. Use common sense regarding safety on the job and comply with any and all applicable OSHA standards.

2. Work with your employer to identify hazards on the job.

3. Report any job-related injury or illness promptly and seek recommended treatment.

4. Follow your employer's safety and health rules and regulations, including the use of personal protective equipment on the job.

5. Identify potential hazards before you begin a task.

6. Respect precautions — don't take chances.

7. Know your company's hazard reporting procedures.

8. Learn basic first aid procedures and how to use them effectively.

9. Report any hazards to a supervisor or designated person as soon as you become aware of the problem.

Personal Safety

1. Be sure you have an approved safety helmet and goggles when you report to the job site.

2. Wear rubber soled safety shoes to help you keep your footing. Open-toed shoes or sandals have absolutely no place on a job site.

3. Don't wear baggy or loose fitting garments. These types of garments can catch on things and tools and cause unnecessary accidents. The same is true of rings and loose jewelry.

4. Don't open up the way for accidents by wearing clothing that is inappropriate for the occasion. Tank tops and shorts should be considered as inappropriate for the work place.

5. Wear a tool pouch. Don't carry tools in your hip pocket or try and climb a ladder with a handful of tools.

6. If your employer does not provide them, invest in a good pair of leather work gloves. There are going to be frequent occasions when you will be handling rough and splinter prone materials. Your hands will also thank you when you wear your gloves to pull wire.

Eye Protection

Under any condition when there might be the possibility of incurring an injury, wear your safety glasses. Any procedure involving power tools used to chip or hammer materials absolutely dictates the use of protective eye glasses.

Employees of system contracting firms are not normally exposed to hazards to the eye that might be more likely in a manufacturing environment involving caustic chemicals or metalworking equipment. However, the hazards of the construction site are very real and the potential of suffering an injury to the eyes is always a concern.

An old safety proverb makes the point extremely well:

> *"You can walk with a wooden leg, you can chew with false teeth, but you can't see with a glass eye"*

Protective eye equipment includes:

Safety glasses. These generally look very similar to ordinary prescription glasses. However, the lenses come in a variety of materials and construction for protection against different hazards.

Goggles. Similar to safety glasses, but afford more protection since they generally have some shielding to the sides and fit closer to the eyes. Models are available that can be worn over prescription glasses.

You should maintain and clean your safety glasses and/or goggles regularly. Dirty, scratched, or cracked lenses reduce vision and seriously reduce protection. Replace damaged glasses immediately.

Hearing Protection

OSHA's regulations covering hearing protection are published under CFR 1910.95.

Exposure to sharp, loud noises (impact noise) or to sustained durations of noise can cause serious and permanent hearing damage.

As employees in the sound and communications business we frequently are exposed to conditions where excessively loud sound levels can be generated. Therefore, we must be twice as careful to avoid exposure to noise and sound levels that could prove damaging to our hearing.

We recognize that there are different types of noise. For the purpose of hearing protection, these are divided into three categories:

Wide band noise is distributed over a wide range of frequencies which we in the sound and communications industry would commonly refer to as white noise.

Narrow band noise is of limited bandwidth and would be associated with equipment such as sanders, circular saws, fans, etc.

Impulse noise is, as the name implies, short duration 'bursts' of either wide, or narrow band noise.

We now know that exposure to noise or sound pressure levels in excess of 85 dBA over an extended period of time can result in cumulative hearing damage. Sharp bursts of high level sound can cause almost instantaneous damage to the ear.

Hearing protection devices are available in three forms of ear plugs (auras) that fit into the ear canal:

Formable plugs are made of waxed cotton or acoustical fiber and are disposable.

Custom-molded plugs are made for a specific individual out of silicone rubber, or plastic molding compound.

Molded inserts are made from soft silicone rubber or plastic. They conform to the user's ear canal and are reusable.

Both types of molded ear plugs must be kept clean with warm, soapy water after each use to avoid ear infections. Hearing protection devices, such as those previously mentioned, will generally have a 25 to 30 dB noise reduction factor.

In industrial and job site situations, the need for hearing protection is frequently posted. When such postings are in place, always utilize hearing protection. The same holds true when using power tools.

Don't risk hearing loss, follow the procedures outlined below:

1. Disposable ear plugs may be more convenient than other types, but to be effective, they need to fit properly.

2. Individuals who are exposed to noise levels of 85 dBA, or more, over an eight hour period are required (by OSHA) to have an annual audiometric test to check their hearing.

3. Keep hearing protectors in good operational order.

4. Don't use homemade hearing protectors such as wadded cotton or paper.

Foot Protection

The OSHA regulations that apply to protective footwear are published at 29 CFR 1910.132 - .136. These regulations, amongst other things, stipulate that safety shoes must meet the requirements of the American National Standards Institute (ANSI) consensus standard on protective footwear (ANSI standard Z41-1991).

The common workplace hazards that can cause foot injuries are:

1. Sharp or heavy objects falling on the foot.

2. The foot or toes are squeezed between two objects or rolled over by a heavy object.

3. A sharp object like a nail breaks through the sole and punctures the foot.

4. Contact with surface hazards like oil, water, or chemicals that cause the worker to slip and fall.

5. Extreme heat or cold which requires insulation or ventilation for the feet.

There are many different kinds of safety shoes available. Some of these are very job specific such as those used by workers handling or working around chemicals, or by foundry workers.

System contractor employees should pay particular attention to shoe types which offer protection from punctures, toe protection and be suitable for climbing ladders and scaffolds.

Make sure your safety shoes meet the ANSI standards.

Hand Protection

Probably the most vulnerable parts of your body for an accident are your hands and fingers. Our hands are exposed to many, many hazards in the workplace. OSHA regulations concerning hand protection are found at 29 CF 1910.132 - .138.

Three general hazards which can cause hand injuries are:

1. Mechanical hazards which are present when using tools, or machinery. Use of such equipment can cause cuts, bruises, punctures, abrasions, or crushing.

2. Environmental hazards like extreme cold, or heat. Exposure to electrical currents or exposure during handling of hot or extremely cold materials.

3. Contact with irritating substances.

As with other types of personal protective equipment, there are any number of hand protection devices, some of which are very job specific. Whatever type you choose, pay attention to the fit and the suitability for the workplace situation where you will be wearing them.

There are few activities on or off the job that don't involve your hands in some way. Driving, eating, writing, holding a loved one — the list could go on and on. Make sure your hands last a lifetime.

Questions

1. What are the hazards of electricity?

    a. Shock.
    b. Burns.
    c. Explosions.
    d. Fires.
    e. All of the above.

2. The single most important rule when lifting is to:

    a. Weigh the load.
    b. Bend the knees.
    c. Lean over to pick it up.
    d. Set the load down properly.

3. All of the following are common sense rules to follow on the job except:

    a. Identify all potential hazards before beginning.
    b. Stick to your normal routine, even if circumstances change.

    c. Check with your supervisor if you are unsure about a situation.

    d. Know your company's hazard reporting procedure.

4. What is the most important aspect of eye protection?

    a. Style.

    b. Durability.

    c. Looks.

    d. Effectiveness.

5. What is the acronym of the organization which sets standards that safety footwear must meet?

    a. ISNA

    b. FEET

    c. TOES

    d. ANSI

6. What is the most used tool you have?

    a. A hammer.

    b. A screwdriver.

    c. A wrench.

    d. Your hands.

7. Hearing protection must be provided when noise intensity exceeds an average of:

    a. 30 dB

    b. 50 dB

    c. 85 dB

    d. 102 dB

(Answers in Appendix A.)

Appendix A
Answers to
Chapter Questions

Chapter 1
COMMUNICATIONS

1. a. Electrical reflected ceiling drawings.
    b. Written specifications.

2. Many consulting engineers will use a standard symbol block regardless of whether or not all the devices depicted on the table are actually incorporated into the project.

3. Document the problem and immediately advise your supervisor of the discrepancy.

4. F.B.O. means "Furnished By Others."

5. If prior to the bid process, clarification should be requested, in writing, to the appropriate specifying authority.

Chapter 2
TECHNICAL FUNDAMENTALS

Operation With Fractions

1. a. 41/48 b. 4 1/30 c. 21/40 d. 2 1/7

2. a. 187/175 b. 46/9 c. 549/500 d. 59/24

3. a. 162.8336 b. 0.0819 c. .8430

4. a. .1325 b. 635.8730 c. 43.5053

5. a. 54.11 b. 763 c. 0.288

6. a. 1 1/3 b. 1/3 c. 4/7

7. a. 1-1/40 b. 3/80 c. 11/40 d. 3 1/10

Percentages

1. The measurement of uncertainty of the instrument is 2% of 500 watts.

    $$\text{wattmeter uncertainty} = \pm(0.02 \times 500) = \pm10W$$

    The measured value will lie between (150 - 10) watts and (150 + 10) watts; 150 ±10 watts.

2. 76.3Ω.

3. 88.9% oxygen; 11.1% hydrogen.

Measurements

1. a.   Absolute error = measured value - handbook value
    Absolute error = 342 m/sec - 346 m/sec = -4 m/sec

    The negative value indicates that the measured value is smaller than the accepted, or handbook value of the speed.

    b. Relative error = absolute error/handbook value:

    $$= \frac{-4 \text{ m/sec}}{346 \text{ m/sec}} = -0.01$$

2. 0.02 for the 0-5 voltmeter; 0.08 for the 0-50 voltmeter:

Exponents and Radicals

1. a. 3^5 b. 5^{-5} c. 16^{-3}

2. a. 2^3 b. 3^{-1} c. $(4p)^0$ or 1

Scientific Notation

1. a. 8.42×10^{11} kw = hr b. 1.793×10^8 people
    c. 4.7×10^3 kw = hr per capita

Ratio and Proportion

1. a. $x = 2$ b. $m = 360$ c. $t = 5.8$

2. The weight of 6'4" of the wire is 0.697 » 0.7 pounds.

3. The tree is 75 feet high.

4. 4.32 amperes

5. 9 cm.

6. 28 teeth

Area and Volumes

1. a. 8.26×10^1 sq.in. b. 5.33×10^2 cm^2, 5.33×10^{-2} m^2

BASIC PHYSICAL SCIENCE

Mechanics

1. 1,425 J

2. ≈ 13,000 J

3. a. Both do the same work. b. Atlas is more powerful.

Acceleration

1. -10 m/sec²

Properties of Electricity and Magnetism

1. Volume and area.

2. Gravity.

3. 1,000 kg.

4. No.

5. 75.8%.

6. ≈ 344 m/sec. or 1130 ft/sec.

7. 0.0688 m. (6.88 cm.)

8. 880 Hz

9. When you double the distance from the source of the energy, the energy density is reduced as to the square of the distance. Or, every time you move away from the source by a factor of two, the intensity of the sound is quartered.

10. The magnetic field is directly proportional to the current flow; hence the magnetic field strength will be doubled.

DC CIRCUITS

Resistors In Series

1. 10Ω

2. a. IR_1 (50Ω) = 500 volt b. IR_2 (10Ω) = 100 volt

3. a. Since the loads are in series the same current flows in both
        loads I = 0.8 A.
    b. The voltage across the 5Ω load is 4 V; the voltage across
        the 10Ω load is 8V.
    c. Load 1 = 3.2 W; load 2 = 6.4 W; the combined load is
        (3.2 + 6.4) = 9.6 W.

4. a. The current in the 2 kΩ resistor = 0.001 A; the current in
        the other therefore also = 0.001 A.
    b. The potential difference across the 5kΩ resistor = 5 V.

Voltage Dividers

1. a. The total of the load currents is 20 mA + 40 mA + 10 mA =
        70 mA. The bleeder current is 10%, so I_1 = .007 A or 7 mA.
        The potential between terminals A & B = 60 V; hence R_1 =
        8571Ω.
    b. The current I_2 is the sum of the 10 mA current to the load at
        terminal B and the bleeder current I_1 (10 mA + 7 mA) = 17
        mA = 0.017 A. The potential between B & C = (V_C - V_B) =
        100 V - 60 V = 40 V; hence R_2 = 2353Ω.
    c. I_3 = 0.057 A, V_{D-C} = 100V, R_3 = 1754Ω

2. a. R_1 = 568W
    b. R_2 = 12,500W
    c. R_3 = 4167W

Resistors In Parallel

1. 3.33Ω

2. Connect a 33.3Ω resistor in parallel.

3. a. R_t when connected in series = 480W, P = 15Ω
    b. In parallel the voltage across each of them = 120V; each bulb dissipates 60W and the lamps will be brighter in parallel.

4. a. I in 20Ω = 3 A. b. I in 40Ω = 1.5 A c. I in 50Ω = 1.2 A
    d. I_t = 5.7 A

Resistors In Series/Parallel Networks

1. R_t = 4.8Ω

2. a. I_1 = 1.25 A b. I_2 = 1.25 A c. I_3 = 2.5 A d. I_4 = 1.67 A
    e. I_t = 4.17 A

Resistance and Wire Size

1. 12.3 feet

2. a. 0.32Ω b. 32Ω

3. 4106 cmil

4. 226.5 volts

5. a. 30.5 ft. b. 122 ft.

Chapter 3
SYSTEM ELEMENTS

Electronic Components

1. a.  b.

2. 500Ω ±10%

3. The farad, often expressed in picofarads (pF), or microfarads (mF).

4. 78 pF ±20%

5. L measured in henrys (h).

6. a. Transformers b. Solenoids c. Relays

7. a. An LC circuit. b. A filter circuit.

8. To *filter* the rectified AC current thereby creating a *smoother* DC current.

Basic Switches and Connectors

1.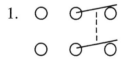

2. Form C

3. Diode

4. The room speaker is switched between the home run pair and the call-in pair.

5. The shield or ground.

6. A 2-circuit phone jack.

7. RG-59U , RG-6U, RG-11U

8. 50Ω

9. BNC

10. The second tip and ring.

Chapter 4
SYSTEM COMPONENTS

1. NEC code book, section 100-A

2. Fire blocks.

3. Line transformer.

4. Tile bridge or rails.

5. Forward, or out away from the basket.

6. Ribbon microphone.

7. 105 degrees.

8. 150 Ω

9. Pin 2.

10. 0VDC

Chapter 5
FIRST AID AND SAFETY

1. e

2. b

3. b

4. d

5. d

6. d

7. c

Appendix B
The NICET Audio Certification Program

The only nationally recognized certification available to installers, technicians, and designers of audio systems is offered by NICET (the National Institute for Certification in Engineering Technologies). NICET certification is a demonstration that the audio technician has competency in standard technical skills and knowledge considered essential to the audio industry. Getting certified by NICET involves meeting a number of requirements, not the least of which is a battery of written examinations. The following section describes what the NICET certification process consists of, and gives suggestions for getting prepared for NICET testing. Those interested in certification should definitely contact NICET directly to request specific program details and applications. NICET may be contacted at the following address:

NICET
1420 King Street
Alexandria, VA 22314-2715
Call Toll-free: (888) 476-4238

Levels, Work Elements and Requirements

NICET currently offers three levels of certification in the field of audio. Level I is intended for the entry-level or relatively inexperienced audio installation technician. To be certified at level I it is not necessary to have any documented work experience or post-high school education. However, some familiarity with basic electronics and audio components is expected. Level II certification requires at least two years of experience in installation, maintenance or design-related work, with at least one year of experience in audio. Also required is either a two-year (associate's) degree or the equivalent in on-the-job experience. Level III is much like level II, except the requirement is that candidates have at least three years of communications systems experience, with at least two years specifically in audio. While a level IV certification has been planned, it is currently not available.

If you can prove that you meet the work experience requirements in your application to NICET, the next step is to take written exams which NICET schedules at colleges, schools and universities all throughout the USA four times a year. Each certification level has its own unique set of subject matter areas, which are referred to as "work elements." The work elements consist of sets of multiple-choice questions on a specific topic. The work elements are categorized in one of three ways: "general core" work elements, "general non-core" work elements, and "special" work elements. Why are there different kinds? The NICET exams are intended to be flexible. Not everyone knows or uses the same information to do audio installation and design work. NICET expects each technician to know some basic, universal things, but recognizes that some knowledge and skills differ from person to person depending on where they work, or what his or her job responsibilities are. The basic, universal knowledge is what is considered "core." The non-core work elements apply to most technicians, but when applying to test, only a fixed number of the total available must be passed. As the examinee, it is up to you to choose which of these non-core elements you want to attempt. This is also true of the "special" work elements, which cover special kinds of systems and skills which are not universally practiced by audio technicians, but with which you may have experience. The NICET "Program Detail Manual" (available from NICET) describes in detail what the work elements consist of and how many of each must be passed to achieve each level of certification.

Applying and Preparing for Testing

Before you begin in earnest to study for the NICET exams it's a good idea to get familiar with the application process, rules, procedures, and requirements as outlined by NICET. Make sure you read ALL of the NICET Program Detail Manual as well as the following sections of NICET's "General Information Booklet – Tenth Edition": Section I – General Information; Section III – Technician General Information; Section IV – Work Element Exams; Section VII – Operational Policies. As mentioned earlier, you must apply to NICET to register for tests. The application consists of a form much like a job application where you must supply information about your work and educational history, address, etc. In addition, you must have your supervisor or someone who is familiar with your work verify your experience in the areas on which you plan to be tested. The other part of the application is a personal recommendation form where someone you know attests to your positive professional and personal attributes.

A Personal Study/Review Program

It is up to you whether or not to study for the NICET exams. Some technicians are very experienced and have had a lot of training throughout the years, and they can take and pass the NICET exams without any preparation (especially level I). This is not recommended, however. Even though the NICET tests are intended to measure whether or not you possess minimum standards of skill and knowledge in technical audio, it is important to realize that the standards are very high. This is what makes being NICET-certified truly meaningful — not just anybody can pick up a test and pass it. In most cases, it is a very good idea to prepare for the exams.

In addition to this book and the others in the NSCA Audio Technology series, there are several resources that you might want to use in your own study program. First, on pages 24 and 25 of the NICET Program Detail Manual you will find a list of books that contain a lot of information about audio and related systems. Most of these are available at your local bookstore, or from NSCA's On-line Bookstore (HYPERLINK http://www.nsca.org). Many of these titles are considered the industry "bibles" that just about every professional audio person has read or referred to at one time or another. It is a must to have a current copy of the National Electrical Code((NEC), and you are also strongly encouraged to get your hands on a good math textbook and electronic theory book. You can take many of these items to your NICET exams (see the NICET policies concerning testing) but don't plan on blowing off your studies thinking you'll just take a whole library of books to the test. Even if you had every book in the world to refer to at test time, there's simply no way you would ever have enough time to look everything up and answer all the questions. If you plan to take any reference books to the test, your best bets are this book (and the other books in the series) and the National Electrical Code (NEC). Part of your preparation time would be well spent highlighting tables and facts in these books, especially things that are difficult to memorize, or things in which you are not so well-versed.

Besides reference books, you will also want to take a good scientific calculator to the test. Part of your personal study program should involve practicing with the calculator you plan to use. Get familiar with its functions and how its notation translates into other forms of notation you may find in math textbooks or audio books. Whatever you do, make sure you take a calculator you know how to use.

An excellent way to jump start your preparation for the tests is to attend one of NSCA's CATT Seminars (Certified Audio Technician Training Seminars). CATTS are three-day seminars that provide a structured review of the most difficult and important work elements you will encounter on the NICET tests. CATT Seminars are offered about once a month in various cities throughout the US. Call NSCA at (800) 446-6722 or (319) 366-6722 for more information. You can also visit the NSCA web site (HYPERLINK http://www.nsca.org) for updated information on CATTS.

Start your study program by carefully reading the NICET Program Detail Manual. Get familiar with all of the work elements, especially those in levels I and II, since that's where your testing will begin. If you're like most technicians, you probably are stronger in some topic areas than others. In that case, it would be overkill to study for each work element with equal fervor. Narrow the focus of your studies down by making a list of those work elements that concern you the most. Conversely, decide which work elements you feel reasonably confident about and leave your review of those for the end of your study effort. You may find that you've run out of time to prepare (The test is tomorrow!) and you'll be better off having spent time on the things you are the rustiest in, rather than just going down the list more or less arbitrarily.

Do your homework. Once you've established which work elements you need to brush up on the most, decide what your study resources for that topic will be. Make a list of books, people you can ask, videos, or even chapters in this book that you will get information from. Do it methodically. Tests are methodical, and they are made by methodical people. So if you want to be successful, you too have to think methodically.

After you've decided on the work elements you will focus on and figured out what you will use to study from for each, then figure out how much total time you have until the test. Decide how many days or weeks you are going to spend studying each topic. You don't necessarily want to divide up your time equally. It depends on what topics you need the most work on. This is the time to get out a calendar and WRITE IT DOWN. This many weeks for this, this many weeks for that, etc. Set aside a fixed time of day, or day of the week to study. Think of your study time as part of your job: if you miss you might be in trouble. Reduce the temptation to procrastinate or to be distracted by making sure others, such as family members and friends, understand that your study time cannot be put off.

As it was pointed out earlier, you won't have time at the test to pore through books looking for the answers to every question. For this reason, it is important to mark the books you will bring with highlighters, "sticky notes," or bookmarks. You'll have to decide what to mark, but things like formulas, reference tables, and other lists of things that you don't normally need to memorize are the most likely candidates. Get familiar with where things are in your books, and make it easy to go right to them.

Your final preparation the evening before, and the morning of the test should involve rest and sensible eating and drinking. Don't stay up all night "cramming" for the test and drinking five gallons of coffee. In most cases you'd be better off not studying at all and going to bed early.

Test-Taking Strategies

NICET's tests are multiple-choice. However, don't be fooled into thinking that for each question there will always be one answer that is obviously correct while the rest are obviously incorrect. In many cases the differences between one possible answer and the next are very subtle. Also, since your total test time is based on the number of work elements and questions you signed up for, you should not spend too much time on each answer. A good strategy is to give yourself a minute or less to answer each question. If you still haven't answered the question after one minute, move on to the next question, and plan to go back if you have time. You are more likely to get more answers correct overall this way. If you are running out of time, or if you just can't answer some questions no matter how hard you try, do not leave any question blank. You are better off guessing or putting down a random response than putting nothing at all. Statistically, even if you randomly answered all of the questions, there is a good probability that you'd get approximately twenty percent or more correct. It's not so different than rolling dice. Now twenty percent won't allow you to pass, but if you intelligently answered most of the questions, then the twenty percent on the rest is better than zero, and it could make the difference between passing and not passing.

Moving Up: Strategies for Subsequent Tests

If you plan to get certified at a level II or III, then you should sign up for the maximum allowable work elements on your first test, which is thirty four. This should include at least twelve level I work elements (the number needed for level II certification) with the remaining twenty two coming from the level II work elements. This MUST include all of the core work elements. See page 3 of the NICET Program Detail Manual for the specifics on which work elements are required for each level of certification. Even if you only want a level I, you should sign up for at least twelve, if not all fourteen, of the level I work elements just in case you don't pass one or two. Nine must be passed to get a level I certification.

If you are signing up for your second or third exam sitting, you should make sure and sign up to retest on any work elements you failed, then sign up for any core elements that are required, and finally, choose which non-core general and special work elements you feel the most comfortable with. Plan on going in for testing at least twice if you want a level II, and at least three times if you want a level III. If you fail a work element, you can take it again (though the questions will be slightly different) twice before you will be required to get remedial training.

Costs of Certification

As of this writing, the cost at application time to take a NICET test is $90.00 (US). You will have to pay this amount each time you sign up to test. If for some reason you can't make the scheduled test time, and if you give at least nineteen days advance notice, you can reschedule without reapplying, and without paying an additional fee. If you give from five to eighteen days advance notice, you can reschedule without reapplying, but must pay an additional $45.00.

See the NICET General Information Handbook for more information on other fees for other circumstances.

Recertification

Once you are certified by NICET your certified status is good for three years. At the end of the three-year period you must get recertified. This is accomplished by providing to NICET evidence of efforts you have made to improve your skills and knowledge through continuing professional development. NICET awards Continuing Professional Development Points, or CPDs based on work experience and training. Most of your points can be earned simply by continuing to work in the audio field. Other ways to earn points are by taking technical, management, and other professional development courses at colleges, training seminars, technical schools, and the like. Earning CPD points is not difficult, as long as the classes or training you take cover skills or knowledge that haven't already been counted toward your existing certification or previous CPD points. See policy # 30 on pages 59-63 of the NICET General Information Handbook for the specific details about how CPD points are earned and applied to recertification.

Being Certified: Promote It

Achieving certification is something to be proud of! It's also a tool for promoting yourself and your employer. After you get formal notice of certification, which includes a certificate and an official NICET reference number, take advantage of the status by attaching your new credential to your name on business cards, company letterhead, correspondence, and your resume. For example, if you are "John Doe" before you are certified, you'll be "John Doe, CET" once you get certified. The "CET" stands for "Certified Engineering Technician."

You might just think it's a few letters, and that it doesn't do much for you, but the fact is that both employers and customers care about it. It shows to them that they are working with true professionals, and it suggests not only that you know what you are doing and that your work is reliable, but also that you have pride in yourself and the work that you do. Certifications exist in many different areas of today's commercial culture, from auto repair to customer service, from medical technology to public accounting. In many of these industries you wouldn't even consider working with someone who isn't certified. Your certification could be the catalyst for you receiving a pay raise, an advancement, or it could give you entry to a new and better position at a new employer. It can also be used by your employer to win contracts which will make the likelihood of profits being passed on to you even greater.

There is another reason why getting certified may be important for you in the near future: it is predicted that by the year 2001 most contracts for audio installations will require that all work be done by certified technicians. If this is true, then without certification you may be out of a job! Getting certified now has many benefits and is relatively painless for most experienced audio technicians, while getting certified later may be... well, may be too late.

Index